Networking Simulation for Intelligent Transportation Systems

Series Editor
Abdelhamid Mellouk

Networking Simulation for Intelligent Transportation Systems

High Mobile Wireless Nodes

Edited by

Benoit Hilt
Marion Berbineau
Alexey Vinel
Alain Pirovano

iSTE

WILEY

First published 2017 in Great Britain and the United States by ISTE Ltd and John Wiley & Sons, Inc.

ISTE Ltd
27-37 St George's Road
London SW19 4EU
UK

www.iste.co.uk

John Wiley & Sons, Inc.
111 River Street
Hoboken, NJ 07030
USA

www.wiley.com

Library of Congress Control Number: 2017930998

British Library Cataloguing-in-Publication Data
A CIP record for this book is available from the British Library
ISBN 978-1-84821-853-6

Contents

Preface xi

Chapter 1. Simulation of Convergent Networks for Intelligent Transport Systems with VSimRTI 1

Robert PROTZMANN, Björn SCHÜNEMANN and Ilja RADUSCH

1.1. Introduction 1
1.2. Fundamentals of cooperative ITS 2
1.2.1. Message types 2
1.2.2. Application categories 3
1.2.3. Supporting facilities 4
1.3. Overall simulation framework 5
1.4. Simulation of cellular networks 6
1.4.1. Regions and cells 10
1.4.2. Delay models 11
1.4.3. PR-Model and PL-Model 12
1.4.4. Capacity Model 13
1.4.5. Topological and geographical messaging 14
1.5. Simulation study 14
1.5.1. Evaluation metrics 16
1.5.2. Simulation set-up 18
1.5.3. Simulation results 21
1.6. Conclusion 25
1.7. Bibliography 26

Chapter 2. Near-field Wireless Communications and their Role in Next Generation Transport Infrastructures: an Overview of Modelling Techniques 29

Christian PINEDO, Marina AGUADO, Lara RODRIGUEZ, Iñigo ADIN, Jaizki MENDIZABAL and Guillermo BISTUÉ

2.1. Near-field wireless technologies 30
2.1.1. Near-field versus far-field 30
2.1.2. Near-field-based technologies in transport 33
2.2. Characterization of near-field communications 36
2.2.1. Electrical models 37
2.2.2. Analysis of the mutual inductance of a squared inductive coupling 37
2.2.3. Computer-aided electromagnetic calculation 40
2.3. Discrete event simulators 42
2.3.1. Riverbed Modeler 43
2.3.2. OMNeT++ 44
2.3.3. ns-2 45
2.3.4. ns-3 45
2.3.5. Discrete event simulator comparison for near-field communication 46
2.4. Conclusions 47
2.5. Bibliography 48

Chapter 3. Trace Extraction for Mobility in Civil Aeronautical Communication Networks Simulation 51

Fabien GARCIA and Mickaël ROYER

3.1. Traffic regulations 52
3.1.1. General airspace 52
3.1.2. North Atlantic airspace 53
3.2. Mobility for network simulation 54
3.2.1. Types of mobility models for AANETs 54
3.2.2. Comparison of mobility model types 55
3.3. Example of mobility trace extraction 56
3.3.1. Extraction of information 57
3.3.2. Traces filtering 57
3.3.3. Enhancing traces 58
3.4. Toward cooperative trajectories 60
3.5. Bibliography 60

Chapter 4. Air-Ground Data Link Communications in Air Transport 61

Christophe GUERBER, Alain PIROVANO and José RADZIK

4.1. Introduction 61
4.1.1. Context. 61
4.1.2. OMNeT++. 63
4.2. Continental air-ground data link communications and VDL mode 2 63
4.2.1. Communication system. 63
4.2.2. Dimensioning parameters and bottlenecks 65
4.2.3. Simulation model 67
4.2.4. Analysis of simulation results 69
4.3. Oceanic air-ground data link communications and AMS(R)S. 71
4.3.1. The aeronautical mobile satellite (route) service and Classic Aero 71
4.3.2. Dimensioning parameters and bottlenecks 73
4.3.3. Simulation model 74
4.3.4. Analysis of simulation results 75
4.4. Summary and further work 76
4.5. Bibliography 77

Chapter 5. A Virtual Laboratory as an Assessment Tool for Wireless Technologies in Railway Systems 79

Patrick SONDI, Eric RAMAT and Marion BERBINEAU

5.1. Introduction 80
5.2. ERTMS subsystems and related test beds 81
5.2.1. The functional subsystem of the ERTMS. 81
5.2.2. The telecommunication subsystem of the ERTMS 84
5.3. A virtual laboratory based on co-simulation for ERTMS evaluation 86
5.3.1. Why a co-simulation approach? 86
5.3.2. Which data and processes must be modeled in each simulator? 87
5.3.3. Overall architecture of the ERTMS–OPNET virtual laboratory. 89
5.3.4. Synchronization modes. 90
5.3.5. Virtual laboratory implementations in the ERTMS simulator 92
5.3.6. Virtual laboratory implementations in OPNET 93
5.3.7. Virtual laboratory implementations in the co-simulation manager. 95

5.4. Effective use of the ERTMS–OPNET virtual laboratory 97
5.4.1. A co-simulation scenario with the ERTMS–OPNET virtual laboratory 97
5.4.2. Efficiency of the co-simulation approach in the evaluation of railway systems 101
5.5. Conclusion 104
5.6. Bibliography 105

Chapter 6. Emulating a Realistic VANET Channel in Ns-3 107

Hervé BOEGLEN, Benoit HILT and Frédéric DROUHIN

6.1. Introduction 107
6.2. Influence of the channel propagation model on VANET simulation 107
6.2.1. A realistic IEEE802.11 PHY layer 108
6.2.2. Accurate VANET channel propagation modeling 109
6.3. A way to realistic channel modeling with ns-2 112
6.4. Realistic channel modeling with ns-3 114
6.4.1. The Yans WiFi model 114
6.4.2. The Physim Wi-Fi model emulating OFDM-based transmission 115
6.4.3. Data transmission at ns-3 PHY level 116
6.4.4. The internals of WiFi channel modeling 117
6.5. Case studies: emulation of realistic VANET channel models in ns-3 117
6.5.1. A simplified VANET channel model for an urban environment 118
6.5.2. A normalized VANET channel model for urban environments 121
6.6. Conclusion and discussion 123
6.7. Appendix A: The Abbas et al. Model Implementation 125
6.8. Bibliography 130

Chapter 7. CONVAS: Connected Vehicle Assessment System for Realistic Co-simulation of Traffic and Communications 133

Justinian ROSCA, Ines UGALDE, Praprut SONGCHITRUKSA and Srinivasa SUNKARI

7.1. Introduction 133
7.2. Related work 135
7.3. CONVAS co-simulation platform 138
7.4. Realistic DSRC channel models 139
7.4.1. CONVAS propagation models 141
7.4.2. Model tuning based on real-world data 142

7.5. Channel model tuning 143
7.5.1. Michigan safety pilot model deployment data 143
7.5.2. Estimation of PDR 144
7.5.3. Model tuning 146
7.6. Connected vehicle applications 149
7.6.1. Intelligent dilemma zone avoidance 149
7.6.2. IDZA implementation in CONVAS 150
7.6.3. IDZA performance criteria 151
7.7. Experimental results 151
7.7.1. CONVAS setup 151
7.7.2. Co-simulation results 152
7.8. Conclusions 159
7.9. Acknowledgments 160
7.10. Bibliography 161

Chapter 8. Highway Road Traffic Modeling for ITS Simulation 165

Marco GRAMAGLIA, Marco FIORE, Maria CALDERON, Oscar TRULLOLS-CRUCES and Diala NABOULSI

8.1. Introduction 165
8.2. Road traffic models 166
8.2.1. Traffic input feeds 168
8.2.2. Mobility models 169
8.3. Fine-tuned measurement-based model 170
8.4. Comparative analysis of road traffic models 174
8.4.1. Case study scenarios 174
8.4.2. Connectivity metrics 175
8.4.3. Results 176
8.5. Fundamental properties of highway vehicular networks 178
8.6. Discussion and conclusions 181
8.7. Bibliography 182

Chapter 9. F-ETX: A Metric Designed for Vehicular Networks 185

Sébastien BINDEL, Benoit HILT and Serge CHAUMETTE

9.1. Introduction 185
9.2. Link quality estimators 187
9.2.1. Hardware-based LQE 188
9.2.2. Software-based 189
9.2.3. Discussion 190

9.3. Analysis of legacy estimation techniques 190
9.3.1. Type of window 191
9.3.2. Window analysis 193
9.4. The F-ETX metric 195
9.4.1. Window management algorithms 195
9.4.2. Multi-assessment approach 197
9.4.3. Routing integration framework 199
9.5. Simulation settings 201
9.5.1. First scenario 202
9.5.2. Second scenario 202
9.6. Simulation results 202
9.6.1. Performance of the multi-estimators 203
9.6.2. Performance of routing protocols 206
9.7. Conclusion 208
9.8. Bibliography 209

Chapter 10. Autonomic Computing and VANETs: Simulation of a QoS-based Communication Model 211

Nader MBAREK, Wahabou ABDOU and Benoît DARTIES

10.1. Introduction 211
10.2. Autonomic Computing within VANETs 212
10.2.1. Autonomic Computing 212
10.2.2. Autonomic vehicular communications 213
10.3. Broadcasting protocols for VANETs 213
10.3.1. Deterministic methods 215
10.3.2. Stochastic methods 216
10.4. Autonomic broadcasting within VANETs 218
10.4.1. Optimization of broadcasting protocols in VANETs 218
10.4.2. Self-management architecture 219
10.4.3. QoS-based broadcasting 221
10.5. Simulation of a QoS-based communication model 222
10.5.1. ADM: autonomic dissemination method 222
10.5.2. Simulation environment 228
10.5.3. Performance evaluation 229
10.6. Conclusion 231
10.7. Bibliography 232

List of Authors 235

Index 239

Preface

Nowadays, network simulation has become more affordable than real-world experiments and the least-expensive mean for the evaluation of networking propositions for Intelligent Transportation Systems. This requires that, for purposes of accuracy, simulation software adapts to the simulated field. Which, for the case of ITS, results in integration of realistic mobility, wireless communication environments, and protocol mechanisms that are as precise as possible.

However, every simulation user should be aware of the fact that simulation only represents the functioning of the real world in a limited way.

In this book, we show how simulation can be used in several domains of ITS, ranging from vehicular to railway and aircraft communication networks, with appropriate examples. In the 10 chapters of this book, several levels of the communication models and the technologies of ITS communication are addressed. This ranges from channel modeling to traffic generation, including access layer and routing.

In Chapter 1, Robert Proztmann *et al.* address the scalability of vehicular communication technologies on the basis of IEEE802.11p when mixed with LTE technology. They present a multi-aspect simulation environment called VSimRTI, a comprehensive framework that connects various simulation tools together to cover all aspects needed for a proper evaluation of new cooperative mobility solutions for ITS.

In Chapter 2, Christian Pinedo *et al.* address the challenges associated with the interaction of the Internet of Things (IoT) and the ITS domain. They aim to provide guidelines on modeling these smart, low-cost, near-field wireless objects and on how to integrate their behavior in traditional network Discrete Event Simulation (DES) tools.

In Chapter 3, Fabien Garcia *et al.* analyze the current traffic regulations in different airspaces. They lay out the constraints in aircraft movement as well as the different types of mobility models and their respective merits. They finally present traffic traces' extraction, enhancement and filtering, leading to new developments on cooperative trajectory studies as a new trend.

In Chapter 4, Christophe Guerber *et al.* deal with data exchanges between on-board and ground systems. They explain how simulation can be a solution to assess the performances of aeronautical communication architectures and protocols through the examples of communication technologies such as VHF Data Link (VDL) and Aeronautical Mobile-Satellite Service (AMSS).

In Chapter 5, Patrick Sondi *et al.* propose, in the context of the European Rail Traffic Management System (ERTMS), a virtual laboratory based on co-simulation. It relies on two existing tools: an ERTMS simulator implementing the functional subsystem (ETCS) and an OPNET simulator that enables the modeling of the whole telecommunication subsystem, namely the GSM-R (Global System for Mobile Communications̃ Railways). They also address the evolution from co-simulation to multi-modeling in order to directly connect the models and avoid the problems related to heterogeneity of simulators.

In Chapter 6, Herve Boeglen *et al.* show the effects encountered when WiFi frames are transmitted over the air. They provide a channel simulation solution, which is a trade-off between computing time and realism. The source code for ns-3 of this solution is provided in an appendix.

In Chapter 7, Justinian Rosca *et al.* present a platform that flexibly integrates a traffic simulator with a communication simulator, thus providing an ideal platform for co-simulating transportation system applications. The communication models can be tuned on the basis of real-world measurements in scenarios such as urban, residential and highway traffic.

In Chapter 8, Marco Gramaglia *et al.* focus on the representation of road traffic for the simulation of highway vehicular networks based on V2V communication technologies and present an original, fine-tuned, measurement-based mobility model.

In Chapter 9, Sebastien Bindel *et al.* explore the Link Quality Estimators (LQE) in the context of VANET. They propose a metric (F-ETX) that automatically adapts to the link quality and provides a trade-off between the dynamicity and accuracy of Link Quality assessment.

In Chapter 10, Nader Mbarek *et al.* show how to adapt the Autonomic Computing paradigm to ITS and in particular to Vehicular Ad hoc Networks (VANETs) in order to enhance the performance of communications in such changing environments. The design of a QoS-based broadcasting protocol is presented as a usage case.

We hope that this multi-purpose book will help the reader to move a step forward in their understanding and/or current work in the domain of network simulation for Intelligent Transportation Systems.

Benoit HILT
Marion BERBINEAU
Alexey VINEL
Alain PIROVANO
February 2017

1

Simulation of Convergent Networks for Intelligent Transport Systems with VSimRTI

1.1. Introduction

For the realization of Intelligent Transportation Systems (ITS), ad hoc networks based on IEEE 802.11p have a long history in research. This technology envisions a decentralized information exchange between mobile vehicles, and also with stationary roadside stations to enable communication with central stations in the public data network (i.e. the Internet). This approach offers several advantages such as the direct exploitation of the broadcast characteristics of the radio channel, which is useful for short message broadcasting in the vehicle's vicinity. However, scalability is a big challenge in this approach, due to a limited communication range and a lack of deterministic quality of service (QoS). With the new generations of cellular networks (mobile phone networks), these drawbacks of vehicular ad hoc networks could be overcome. Cellular networks, e.g. 5G, are emerging as a capable solution not only for mobile Internet services, but also for ITS-specific traffic safety and efficiency matters. Cellular networks exhibit the major advantage of a nearly unlimited communication range, due to their architecture, with only a short wireless part between the mobile device and the base station, and the wired part through the backbone. However, this architecture introduces a particular delay overhead, which makes meeting the strong requirements of many safety applications questionable. A solution could be an intelligent combination of vehicular ad hoc networks and cellular networks to link the advantages of both approaches.

The multi-aspect simulation environment VSimRTI [SCH 11] is a comprehensive framework that connects various simulation tools together to cover all aspects needed

Chapter written by Robert PROTZMANN, Björn SCHÜNEMANN and Ilja RADUSCH.

for a proper evaluation of new cooperative mobility solutions for Intelligent Transportation Systems (ITS). Vehicle movements and sophisticated communication technologies can be modeled in detail. VSimRTI couples different simulators to allow for the simulation of various aspects of future ITS. In the following sections, we describe how we have extended the VSimRTI architecture to enable the simulation of cellular networks. Consequently, we have developed the novel cellular communication simulator VSimRTI_Cell that introduces a grade of abstraction of cellular networks. The developed simulation tool is lightweight and fast enough for larger scale scenarios. However, particularly from the vehicular application perspective, the simulator models important features which are not considered in other related frameworks [PRO 14a, PRO 14b]. Moreover, the new extended VSimRTI architecture not only allows for the analysis of vehicle networks based on cellular communication, but also novel hybrid solutions that combine ad hoc and cellular communication in an intelligent way.

This chapter is structured as follows. In section 1.2, we resume the fundamentals of the system of cooperative vehicles, such as message types, application categories and the specific concept of facilities. Then, section 1.4 introduces the new cellular simulator VSimRTI_Cell in closer detail. In section 1.5, we perform a short simulation study on generic safety and efficiency applications to present the individual advantages of ad hoc and cellular communication as well as a hybrid approach in converging networks in the context of ITS. Finally, section 1.6 concludes this chapter.

1.2. Fundamentals of cooperative ITS

1.2.1. *Message types*

The information exchange in ad hoc networks among vehicles, and among vehicles and infrastructure units is standardized to guarantee interoperability. The two most important message types are the Cooperative Awareness Message (CAM) [ETS 14a] and the Decentralized Environmental Notification Message (DENM) [ETS 14b].

Cooperative Awareness Messages (CAMs) are distributed within the ad hoc network, and provide information of presence, position and the basic status of a vehicle to neighboring vehicles that are located within a single-hop distance. Vehicles generate, send and receive CAMs, as long as they participate in the ad hoc network. By receiving CAMs, vehicles are aware of other vehicles in their vicinity and are informed about their positions, movements, basic attributes and basic sensor information. CAMs are generated and sent by a vehicle periodically.

Decentralized Environmental Notification Messages (DENMs) are used to alert road users to a detected dangerous situation, e.g. a hazardous location, roadworks or a risk of collision with another vehicle. In general, the processing procedure of sending a DENM is as follows: after the detection of a dangerous event, the vehicle immediately broadcasts a DENM to other vehicles which are concerned by the event and are located within the same geographical area. The transmission of the DENM is repeated with a certain frequency and persists as long as the event is present. According to the type of event detected, the DENM is relayed by other vehicles. The termination of the repeated DENM broadcasting is either achieved automatically once the event disappears, after a predefined expiry time, or by a vehicle that generates a special DENM to communicate that the event has disappeared. A vehicle, which receives a DENM, processes the information and, if the information in the DENM is relevant for the driver, it presents an appropriate warning or information on the vehicle's HMI (Human Machine Interface).

1.2.2. *Application categories*

Enhancing vehicle safety and improving traffic efficiency are the two most important aims of vehicular networks. Moreover, communication capabilities in vehicles also allows popular digital services to be provided to the users. The ETSI [ETS 09, ETS 10] and the Car2Car Communication Consortium Manifesto [CAR 07] define several scenarios and use cases for these objectives. The following section gives a brief overview of how vehicular networks are used to share information to advance vehicle safety, increase traffic efficiency or enable comfort applications.

1.2.2.1. *Traffic safety applications*

Vehicular safety applications are characterized, in general, by vehicular communication which is used to mitigate the occurrence of dangerous situations and accidents. Applications, installed in a vehicle, monitor the vehicle's state and the activities of the driver. Relevant pieces of information are transmitted after a relevance check to vehicles in the vicinity. For example, information about the position and speed of a vehicle via CAM or about dangerous locations on the roadway is transmitted via DENM. The received information is used by the safety applications in the vehicle to either inform the vehicle driver or automatically optimize the safety systems for the best possible reaction to a dangerous situation [SCH 11].

For improved vehicle safety, a *Cooperative Awareness* (CA) application and a *Road Hazard Warning* (RHW) application are specified. The CA application warns a vehicle driver if an emergency vehicle, a motorcycle, or a slow driving vehicle is approaching or if a vehicle runs the risk of a collision at an intersection. This application uses the information of the periodically broadcast CAMs for its detections. The RHW application informs drivers about hazardous locations in their close vicinity, e.g. about vehicles driving in the wrong direction, about accidents, roadworks or signal violations. Here, DENMs are used to disseminate information about the dangerous situations.

1.2.2.2. *Traffic efficiency applications*

By exchanging traffic-related information among vehicles and traffic infrastructure units, vehicular traffic efficiency applications improve the efficiency of the transportation network. The received information is analyzed and used, for example, to inform the driver about delays to be expected and to optimize the vehicle's speed and route depending on the traffic conditions [SCH 11].

For an improvement in traffic efficiency, the basic set of applications defined by the ETSI [ETS 10] proposes a *Cooperative Speed Management* (CSM) application and a *Cooperative Navigation* (CoNa) application. The CSM application aims to optimize the vehicle's speed for a better traffic flow. Thus, the application provides either regulatory speed limit information or transmits information necessary for an optimal speed calculation by vehicles at specific road segments or at intersections. Thus, a vehicle can optimize, for example, its speed to reach a traffic light system during the green signal phase. The CoNa application provides services and information, e.g. about the current traffic situation, to allow the vehicles to optimize their travel routes. This application offers a recommended itinerary based on traffic information, enhanced route guidance and navigation, as well as a limited access warning and detour notification.

1.2.2.3. *Comfort applications*

Comfort or infotainment applications are not directly related to the vehicles' mobility, but are part of today's digital lifestyle. This group includes applications like e-mailing, browsing or media streaming. An important aspect of this group is that these applications do not necessarily rely on cooperative M2M information exchange. They are mostly realized on an individual basis and should be evaluated individually. Hence, the evaluation in the later sections will not consider these applications.

1.2.3. *Supporting facilities*

The Facilities Layer is essential to implement vehicular applications in vehicles. It is a sublayer of the Application Layer and provides generic support facilities to the applications. All facilities are classified into three main categories: application support, information support and communication support [ETS 09, ETS 10]: Application support facilities provide common support functionalities for the applications, e.g. station lifecycle management, automatic services discovery, download and initialization of new services and HMI generic capabilities. Furthermore, CAM and DENM management belong to this category. Communication support facilities comprise services for communication and session management, for example the addressing mode and the session support. Information support facilities provide common data and database management functionalities for the applications. An example of an information support facility is the Local Dynamic Map (LDM).

The Local Dynamic Map (LDM) is a conceptual data store which contains topographical, positional and status information within a surrounding geographic area [ETS 14c]. It is relevant to the safe and successful operation of applications. Data can be received from a range of different sources, e.g. on-board sensors, neighboring vehicles, infrastructure units and traffic centers. Thus, the LDM is able to provide information on the surrounding traffic and RSU infrastructure to all applications that require it.

1.3. Overall simulation framework

The assessment of new solutions for Intelligent Transportation Systems is a challenging task. The Vehicle-2-X Simulation Runtime Infrastructure VSimRTI enables the evaluation of collaborative mobility applications and the assessment of new autonomous and cooperative functions of conventional and electric vehicles. VSimRTI connects various simulation tools together to cover all aspects needed for a proper evaluation of new cooperative mobility applications and Advanced Driver Assistance Systems. VSimRTI facilitates the generation of realistic large-scale synthetic probe data for algorithm validation and system testing [PRO 11, WED 09, QUE 08]. Moreover, VSimRTI enables the analysis of elastic mobility scenarios where drivers, traffic infrastructure and cloud services are joined together into one collaborative network.

The aim of the VSimRTI project is to make the preparation and execution of simulations as easy as possible for users. All management tasks, such as synchronization, interaction and lifecycle management, are handled completely by VSimRTI (see Figure 1.1). Several optimization techniques, such as optimistic synchronization, enable high performance simulations [NAU 09]. Special ITS features, e.g. traffic infrastructure units, charging stations and the CAM and DENM message types, introduced in section 1.2, are supported by VSimRTI. Moreover, the various configuration options and comprehensive user documentation assure a high usability.

In contrast to existing fixed simulator couplings, the VSimRTI simulation infrastructure makes the easy integration and exchange of simulators possible [SCH 11]. Thus, the high flexibility of VSimRTI enables the coupling of the most appropriate simulators for a realistic presentation of vehicle traffic, electric mobility, wireless communication and the execution of mobility applications. Depending on the specific requirements of a simulation scenario, the most relevant simulators can be used.

VSimRTI uses an ambassador concept inspired by some fundamental concepts of the High Level Architecture (HLA) [IEE 10]. Thus, it is possible to couple arbitrary simulation systems with a remote control interface. Attaching an additional simulator only requires that the ambassador interface is implemented. For immediate use, a set of simulators is already coupled with VSimRTI. For example, the traffic simulators SUMO [KRA 12] and PHABMACS, the communication simulators OMNeT++ [VAR 08] and ns-3 [HEN 08], the cellular network simulator VSimRTI_Cell, the application simulator VSimRTI_App, and several visualization and analysis tools are prepared for VSimRTI. Figure 1.1 shows a typical simulation set-up implemented with VSimRTI.

VSimRTI has been used by various automotive companies and research institutes to evaluate collaborative mobility applications.

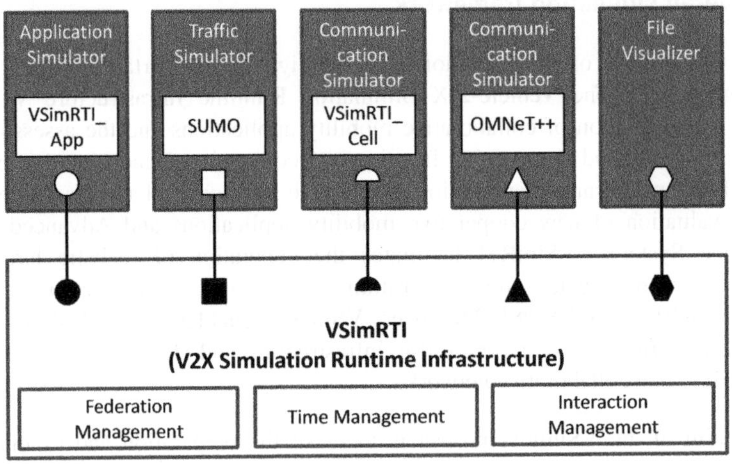

Figure 1.1. *Structure of a typical VSimRTI simulation set-up*

1.4. Simulation of cellular networks

Cellular networks are comprehensive systems with a high number of entities. Moreover, these networks offer very extensive configuration opportunities to match the requirements of the relevant operator. These facts lead to very different characteristics of the particular systems. Hence, the simulation of cellular networks from the perspective of the applications is a challenging task.

The simulation of cellular networks is commonly divided into two different perspectives which have different stages of abstraction. On the one hand, the link level simulation comprises the lower layers (MAC, PHY) and the radio channel. In this way, it models, for instance, the radio link between a NodeB and the UE. On the other hand, the system level simulation focuses on the higher layers and is used for the network view. This level considers, for example, a set of NodeBs and the associated UEs.

Nowadays, different system level simulation frameworks are proposed, concentrating on LTE cellular systems. The longest standing open-source LTE system level simulator is based on MATLAB [IKU 10]. In its original version, it is limited to the downlink and does not consider several important features as broadcast. The C++ based framework LTE-Sim is already very feature rich [PIR 11]. It supports uplink, downlink, several schedulers, handover and more. The well-established communication simulator OMNeT++ is used to build up the end-to-end system SimuLTE [VIR 14]. The latter concept is appealing, as OMNeT++ is already coupled to the existing simulation infrastructure VSimRTI. Even though some of these approaches have a detailed model base, they have several shortcomings for larger scale scenarios. The simulators are more or less tied to one access technology, namely LTE. More significantly, while the direct modeling approach is sufficient for simple ad hoc communication, for larger scale scenarios of cellular system simulation, the given simulators are too complex to configure and the detailed simulation is computationally too expensive. In contrast, trace-based cellular simulation is a promising approach that claims to be much faster than system level simulation [GOE 14]. Similar to the empirical radio propagation modeling, the trace-based technique derives models from real-world measurements. Hence, it works without particular assumptions for the network set-up and configuration.

The new simulator VSimRTI_Cell introduces a similar grade of abstraction of cellular networks to the trace-based simulation. The core models are even based on a dedicated measurement campaign. The developed simulation tool is lightweight and fast enough for larger-scale scenarios. However, particularly from the vehicular application perspective, the simulator also models important features that are not regarded in the other frameworks [PRO 14a, PRO 14b]. The conceptual design of the VSimRTI_Cell simulator has the following key aspects:

Technology: VSimRTI_Cell is independent from the current releases of standardized cellular access technologies such as UMTS-HSPA, LTE or even 5G;

Deployment and Coverage: VSimRTI_Cell introduces a very flexible network deployment concept, which ranges from configuring individual cells to regions of equal coverage;

Network Load: VSimRTI_Cell considers the fact that V2X communication has to coexist with data traffic generated by other users (e.g. with smartphones or USB dongles). The simulation only computes the V2X communication;

Features: VSimRTI_Cell provides important functionalities for the specific needs of V2X communication. For instance, the GEO entity provides the functionality for geographic addressing and information exchange. Moreover, the implemented MBMS functionality allows simultaneous broadcasting of messages to all vehicles in a region or cell.

With the named aspects in mind, the following important metrics for network qualification are identified to be collected within an initial measurement campaign. From these metrics, suitable simulation models are developed:

- transmission delays (see section 1.4.2);
- reliability towards packet losses (see section 1.4.3);
- available data rates (see section 1.4.4).

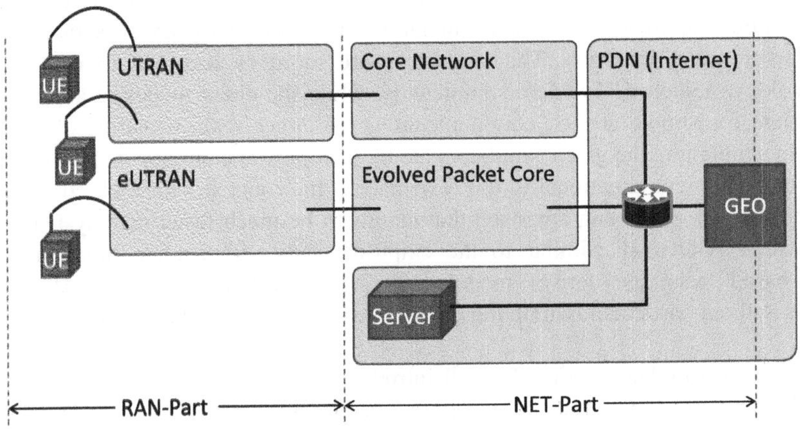

Figure 1.2. *Black box assumption for the cellular system for V2X communication*

The measurement campaign for data collection focused on an end-to-end connection from a smartphone to a server via UMTS. This approach considers the network as a black box, without further assumptions for the specific deployment of the components of NodeBs, RNC, Gateways, etc. in between. Figure 1.2 shows this general assumption for the cellular system for V2X communication. It is based on the established assumption for V2X communication via the central infrastructure. Hence, direct communication uses cases where approaches as D2D are currently not considered. Beside mobile UEs and stationary servers in the PDN, the system also includes a GEO entity, which is introduced for the specific needs of Geographic Messaging in the V2X communication context. The GEO is also located in the PDN. It is explained in closer detail in section 1.4.5. The assumption for the cellular system separates one part for the Radio Access Network (RAN-part) and one part for the Core Network and general public data network (NET-part). The separation intends to enable a more flexible configuration of the overall system.

As the real-world measuring of the communication metrics can be a comprehensive task [GOE 14], the presented concept aims not only to use the data from its own measurement campaign, but also to integrate collected data from others. In this way, the VSimRTI_Cell should also be configured with data from network operators, with measurements from other researchers [SER 09, PRO 09, TEN 10] or with community-driven databases. Several projects such as OpenSignal (www.opensignal.com), RootMetrics (www.rootmetrics.com) and Sensorly (www.sensorly.com) collect crowd-sourced information about the mobile network performance and coverage.

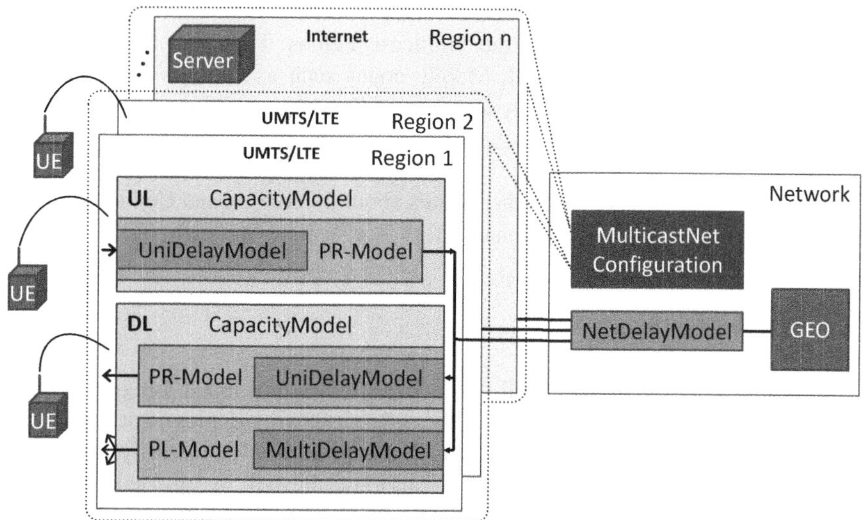

Figure 1.3. *Architecture of the VSimRTI_Cell simulator*

Figure 1.3 shows the architecture of the VSimRTI_Cell. The concept, first, includes multiple regions with specific geographical extensions to create a radio access network with the according coverage properties. Every region consists of one Uplink and one Downlink module to simulate the packet transmission in the RAN-part. In this context, Uplink and Downlink always refer to the direction towards, respectively from, the GEO entity. For instance, a transmission from an Internet-based server towards a vehicle would include an Uplink between the server and the GEO, and a Downlink between the GEO and the vehicle. While the Uplink direction only allows point-to-point communication, the Downlink direction supports point-to-point (Unicast) as well as point-to-multipoint (Multicast) communication. The Uplink module is composed of the three nested models for the Delay, the Packet Retransmission and the Capacity. The Downlink module includes two individual paths for Unicast and Multicast, which share the same Capacity. The Downlink path for Unicast is also composed of the same models for the Delay and the Packet

Retransmission as the Uplink path. The Multicast transmission needs to account for different characteristics. In contrast to reliable ARQ-based Unicast, Multicast only employs FEC with the chance of Packet Losses. Moreover, Multicast typically exhibits a different delay based on the MBMS scheduling period. For this reason, the Downlink Multicast chain provides a separate Delay Model and the Packet Loss Model. All in all, the models for each path (Uplink Unicast, Downlink Unicast and Downlink Multicast) can be individually configured to simulate the according RAN properties.

The second major part of the VSimRTI_Cell models the NET-part. The network enables the configuration of an additional network delay. It furthermore comprises the GEO with its configuration of the Multicast regions. The GEO functionality is implemented in the VSimRTI_Cell. Mobile nodes such as vehicles and stationary servers are the nodes which actually attempt sending and receiving messages. Their application logic is implemented in the VSimRTI_App application simulator.

The following sections give further details about the Region and Cell concept, the transmission models and the functionality for Geographical Messaging.

1.4.1. *Regions and cells*

According to the VSimRTI_Cell design aspects, we developed a region concept that aims at the flexible configuration of the cellular network deployment. In the first instance, regions are independent from actual cells and do not necessarily conform to them. Figure 1.4 shows the possible definitions allowed by this concept. The underlying simulation models allow for the definition of arbitrary polygons as regions. For the sake of simplicity, we decided to present the configuration with rectangular regions, although this would introduce a certain abstraction towards the real-world characteristics:

Free definition (regions ! = cells): this definition typically applies for measured (trace-based) or crowd-sourced data. For instance, the named measurement campaign collected the points for the metrics of the latency, the packet loss and the data rates mainly in connection to their position. The measuring points with equal or similar values are aggregated to the different regions. A further mapping to a certain base station is not performed;

Exact definition (1 region == 1 cell): this definition applies when network operator data about the individual base station positions and their coverage areas are available;

Intra-cell definition (n regions == 1 cell). For more detailed investigations of different coverage areas inside a single cell, the region definition also enables, for example, the configuration of a central region with a more capable parameter set compared to the regions at the cell edges.

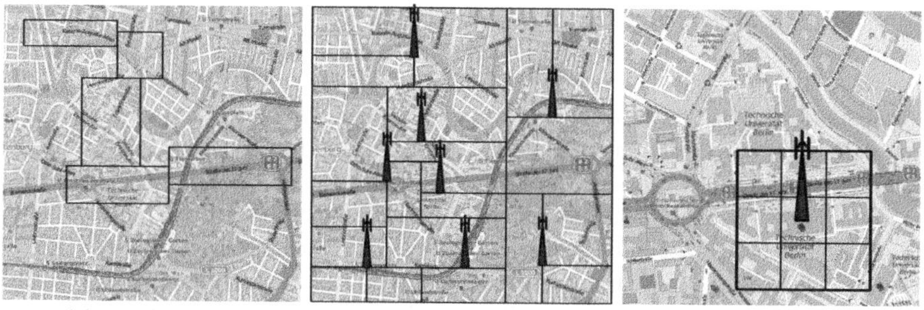

(a) Free definition (b) Exact definition (c) Intra-cell definition

Figure 1.4. *Different definition possibilities for cellular regions in VSimRTI_Cell. For a color version of this figure, see www.iste.co.uk/hilt/transportation.zip*

For practical reasons, the region configurations need to account for two specific situations. First, the whole scenario area may not be covered with a particular region definition, but nodes may move to an uncovered location. In this case, the global region always defines a default configuration. Second, multiple region definitions may be configured to overlap for certain locations. In this case, the configuration of the smallest region is always selected for the transmission calculation.

1.4.2. *Delay models*

The delay models, regardless of the employment as UniDelayModel, MultiDelayModel or NetDelayModel, always constitute the core component for the simulated packet transmission. We developed four different basic delay types to simulate the transmission time for every packet statistically:

constant is the most basic delay type of VSimRTI_Cell. It always yields the same configured delay for every sent packet. This more synthetic model is mainly intended to be used for debugging or primary clarifications. Moreover, it can model a constant offset for the NetDelayModel;

simpleRandom extends the constant delay type. It defines a minimum and maximum bound for the delay ($minDelay$, $maxDelay$) and a possible number of discrete *steps* (n). With this configuration, the simpleRandom type randomly generates n different uniformly distributed delays in the interval of $[minDelay, maxDelay]$;

gammaRandom addresses the particular characteristics of the RAN-part. The measurement campaign identified that the distribution of the transmission delays in a real-world environment sufficiently conforms to the gamma distribution. This delay type allows us to configure the minimum and the expectation value of the delay $(minDelay, expDelay)$;

gammaSpeed is the most sophisticated delay type. It is based on the gammaRandom type and also includes impairments for higher vehicle speeds according to a fitting of the measurements from our campaign. Figure 1.5 displays the probability distribution for the gammaSpeed delay type at different speeds with the measured values of $minDelay = 40\ ms$ and $expDelay = 80\ ms$ for a representative set of HSPA transmissions. According to this diagram, most packets have a delay between 50 ms and 200 ms. However, this is only one possible parameterization and this type also qualifies for the modeling of other mobile network generations such as HSPA+ or LTE and even 5G.

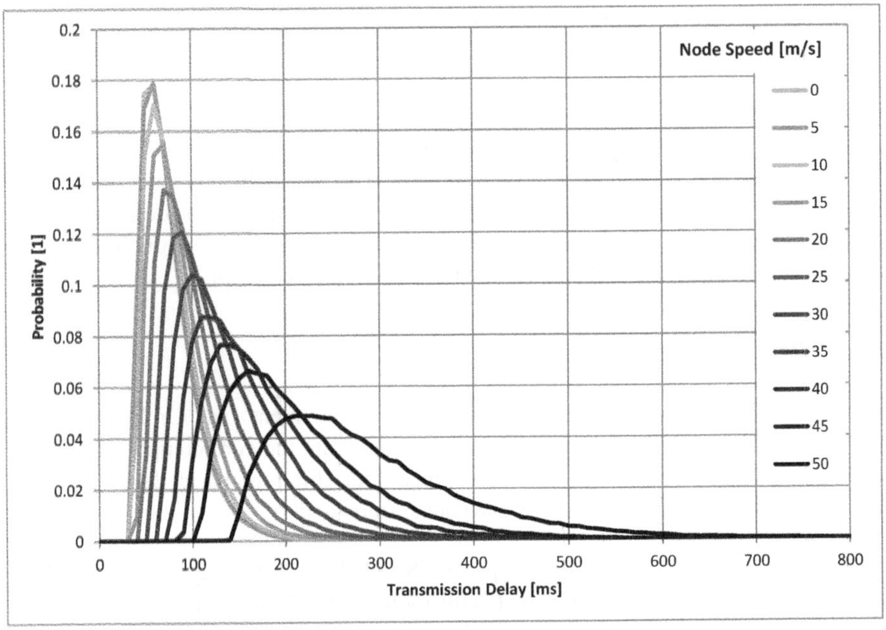

Figure 1.5. *Probability distribution of the gammaSpeed delay at different speeds. For a color version of this figure, see www.iste.co.uk/hilt/transportation.zip*

1.4.3. *PR-Model and PL-Model*

We developed a PR-Model and a PL-Model to address the effect of individual packet transmission impairments between the node and the base station due to inappropriate signal coverage. However, when a reliable connection with ARQ is assumed, no packet is effectively lost, but retransmitted. This is in turn connected with an additional delay. Hence, the Packet Retransmission Model is particularly employed for the reliable Unicast transmissions in Up- and Downlink. For the Broadcast communication in Downlink, where only FEC can be applied, the Packet Loss Model simulates complete packet drops.

The configuration of the coverage quality parameter between 0 and 1 determines the probability of a retransmission (PR-Model) or a packet loss (PL-Model) for each transmission attempt. In case of a packet loss with or without retransmission, the packet will always occupy the channel resources even for unsuccessful transmissions. The parameter value of 0 implies an unimpaired transmission for each model. A value smaller than 1 gives the probability of loss or retransmission in percent. A value of exactly 1 leads to a packet drop in each model. This behavior can be employed to account for entirely disconnected regions in tunnels or shadowed urban canyons. However, the PR-Model optionally reports a packet drop notification to the sender node to consider a reliable transport protocol such as TCP.

1.4.4. *Capacity Model*

Our Capacity Model considers the channel load of a region and calculates the final delay for the individual packets. With the configuration parameter of the maximum available capacity for all simulated nodes, it allows investigations which are independent of the family and the generation of the mobile access technology. Furthermore, it respects static data traffic caused by other mobile users with smartphones, USB dongles and broadband cards. This is an important feature as V2X communication needs to share the resources with other applications. For these reasons, the region definition is particularly important for this model. For example, assume a network deployment with equal capacities in different cells. When this deployment is configured with regions of different size, the capacity needs to be adapted to the region size.

The second parameter of this model is the maximum user bit rate, which resembles the peak speed according to the user data plan. It is still possible to serve more simulated nodes in a certain region than the ratio of the available capacity divided by the maximum user bit rate. When every user demands its maximum bit rate, the result would be that the network gets congested locally and not every sender can transmit directly. This effect is modeled when the sender reserves the resources for the packet at the time of the transmission.

The Capacity Model maintains a resource map where all reservations are accumulated for their timespan. When a new sent packet exceeds either the maxNodeBitrate (the data plan limit is reached) or the available capacity (the network is congested in this region), the packet needs to be queued and thus further delayed until the channel is free again.

1.4.5. *Topological and geographical messaging*

The GEO entity in the Net-part of the VSimRTI_Cell provides functionalities for different addressing schemes. In a real core network deployment, these functionalities would be distributed over several entities, as for instance in LTE, the MME for node mobility management. The GEO is connected to all regions via the NetDelayModel to simulate an additional delay through the Net-part (Core Network and PDN). During simulation runtime, the GEO follows the node mobility. It maintains a table with the node positions and the mapping to the corresponding region. Every sent message in the Uplink goes through the GEO, which distributes the message in Downlink either for point or multipoint reception.

For conventional data traffic, the addressing between the nodes is realized by IP and involves multiple entities in the core network. The simulation can abstract from several aspects of a real core network. However, the user mobility and the router functionality, which are covered by the SGSN in UMTS or the MME and SGW in LTE, need to be accounted for at least. On that account, the GEO uses the knowledge of the current node positions to forward the messages to the Downlink transmission chain of the according region of the destination node. Many V2X communication use cases envision geographic messaging over cellular networks, similar to geographic ad hoc routing. For this purpose, the IP address is extended with the definition of the geographic destination area. The GEO translates the address to direct the packet to the according nodes.

Moreover, many V2X communication use cases demand the dissemination of the same information to multiple nodes in the area. Hence, they are a prime example for the utilization of MBMS and eMBMS (MBSFN) features to allow efficient and resource-saving broadcast transmission. Depending on the MulticastNet configuration, the GEO provides transmission modes similar to MBMS and MBSFN. The MulticastNet configuration defines which regions together form a compound for broadcasting or multicasting a packet. The GEO replicates the packet to be sent in every region compound covered by the destination area.

1.5. Simulation study

The following section presents a simulation study where the introduced cellular simulator VSimRTI_Cell is set into operation. In the study, ad hoc and cellular communication will be combined in one scenario to support the information exchange of V2X applications over converging networks. For a general statement on the communication performance, this simulation will not only address a single application. The evaluation will concentrate on application specific metrics which are significant for a broad spectrum of applications.

As introduced in section 1.2, many envisioned applications rely on the characteristic communication paradigm of periodic exchange of messages [ETS 09, ETS 10]. Hence, the definition of Cooperative Awareness Messages (CAM) is a central point in the specification of the V2X communication standards [ETS 14a]. Additionally, Decentralized Environmental Notification Messages (DENM) represent the second important message type [ETS 14b]. For the properties of node mobility and number of reporting nodes, CAMs and DENMs vary in the manner that CAMs inform about individual moving vehicles, while DENMs inform about (temporarily) stationary situations, which could be reported redundantly by multiple nodes. For the importance of communication reliability, this implies that a CAM possesses more critical requirements, while a lost DENM could be compensated with redundant ones. Due to this, we mainly focus the safety relevant evaluation of our simulation study on applications, which is based on the critical CAMs.

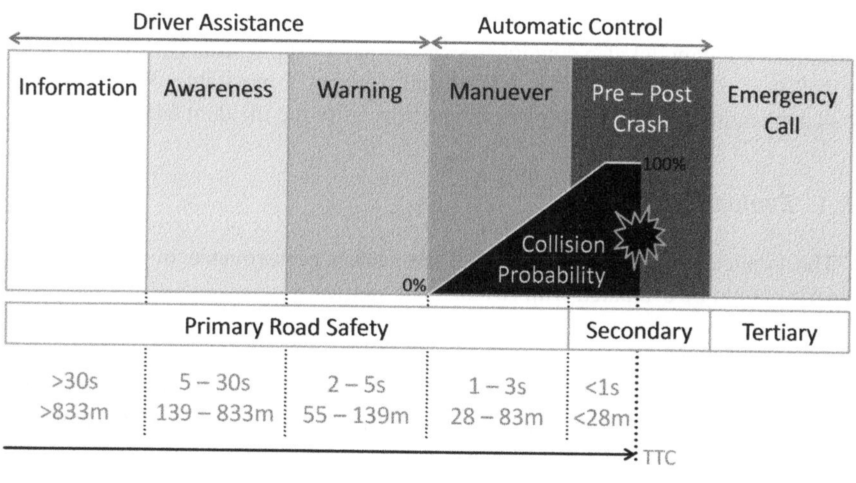

Figure 1.6. *Distance zones of the ETSI road safety application model (based on [ETS 13])*

Figure 1.6 shows the constraints of CAM-based safety and efficiency applications. It shows the position of the different information zones in relation to the time to a possible incident, using the metric of the TTC (time to collision). All values for the TTC should be accepted with caution as exact values are indeed very difficult to define. Even according to the ETSI, the given values are not finalized and are mainly intended as examples [ETS 13]. Figure 1.6 includes additional values for the distance towards the incident to get a better sense of the related spatial dimensions. These values are simply calculated for the TTC of the individual zones, in a situation where two vehicles approach each other with a constant speed of 50 km/h

(13.89 m/s). Different situations (e.g. different movement constellations or vehicle speeds) would obviously lead to other values here.

The leftmost zone of the model contains all applications for driver Information. Such applications have the most relaxed timing requirements in this model and no critical safety relevance, yet the highest distance to the situation. In fact, these applications conform most likely to the traffic efficiency applications from the classification in section 1.2. There is a smooth transition from safety to efficiency applications, as there is also between the individual safety applications with soft and hard timing constraints. The next zones for Awareness, to inform the driver about road hazards, and for Warning, to signal possible collision risks, still contain applications for driver assistance. The Maneuver zone is characterized by an increasing collision probability and a TTC that is below the reaction time of most drivers. Hence, this zone is the last one that contains primary road safety applications to avoid collisions. However, collision avoidance and stabilization would only be possible with the active engagement of the vehicle's automatic control systems. Additionally, the model also contains secondary safety applications in the zone where the collision probability reaches 100 % and a crash is inevitable. Finally, tertiary e-call applications aim for safety relevant actions after the incident takes place.

1.5.1. *Evaluation metrics*

The application and hence the communication performance in the simulation scenario should be evaluated with two distinct metrics.

1.5.1.1. *Safety metric*

For safety use cases, it is particularly relevant that the periodically transmitted information (in CAMs) reaches the destined receivers in time. Conventional approaches to analyze only the packet delivery ratio (PDR), meaning the successfully received messages out of all sent messages, or the transmission latency, meaning the delay from the sending attempt to the reception, deliver only a limited informative value for this issue. The combination of both metrics evaluates the time period between two successfully received messages from an according sender. This metric is known under several synonyms as Consecutive CAM Period (CCP) [PRO 14d], Inter Reception Time [ELB 06], Inter-Packet Gap or Update Delay [KLO 12].

The CCP could be represented with the following equation 1.1, where $n-1$ and n are two subsequently received messages and t_r is the time of reception:

$$CCP(n) = t_r(n) - t_r(n-1) \qquad [1.1]$$

According to this definition, the CCP initially depends on the sending rate f_s and the communication quality, which is actually the property that should be measured.

In the case of ad hoc communication with single-hop broadcast of messages to the neighbors in the communication range, packet losses due to fading or shadowing would lead to an increased CCP compared to the sending rate. Hence, the CCP is qualified to measure burst errors. In the case of communication over a cellular network, packet losses are mitigated by methods of (hybrid) ARQ with message retransmissions on the different layers. However, this approach could result in higher transmission latencies than the sending rate. On the receiver side, this aspect denotes out-of-order delivery for the individual packets. The CCP is also qualified to measure this case, which leads to an increased CCP, as only the most recent updates are useful for the safety applications.

With the given definition, the sole CCP has some minor drawbacks. First, the range of the CCP is in the interval of $[\frac{1}{f_s}, \infty)$. Particularly in cases where the CCP has high values, the potential receiver never receiving updates from the sender, could have two causes. It could either depict critical burst errors. However, the two nodes could also be located far away enough from each other to be anyway out of communication range and thus most probably also out of mutual relevance. Second, the CCP actually measures the supported real-time capability for the certain use cases and the use cases could have very different requirements towards this reaction time.

Hence, the evaluation of the CCP should primarily consider all CCP time spans t_{ccp} where the node i is in the relevance area t_R of a regarded sender and where the t_{ccp} is smaller than or equal to the real-time requirement τ plus a short time difference δ_t. This short time difference accounts for tolerable jitters in the message transmission. The Safe Time Ratio (STR) is the result when this value is normalized with the time span where the nodes are in the relevance area. It is described in (equation 1.2):

$$STR_i(\tau) = \frac{\sum\{t_{ccp}(i)|t_{ccp}(i) \in t_R \wedge t_{ccp}(i) \leq \tau + \delta_t\}}{\sum t_R} \qquad [1.2]$$

The name Safe Time Ratio (STR) was coined in related work [SEG 14]. The definition of the STR shows similarities to the calculation of cumulative distribution function (CDF) of the distribution of the CCP as it considers all measures less than or equal to a specific value. Actually, the complementary CDF is used in the literature for the measurement of unreliable periods [KLO 12].

1.5.1.2. *Efficiency metric*

Due to a higher distance horizon towards the traffic situation, efficiency use cases have more delay-tolerant characteristics. The quality of information reception could be calculated with a mean squared error metric regarding the received information, according to equation 1.3. This metric considers the deviation of the perceived information data $\hat{D}(i)$ at the individual vehicle node i in comparison to the data of the actual reference situation D. For better scalability, the MSE is normalized with the norm of the reference data D:

$$MSE_i = \mathbb{E}\left[\frac{1}{\|D\|^2}\left\|\hat{D}(i) - D\right\|^2\right] \qquad [1.3]$$

For the simulated applications, we use the current speed from the transmitted CAMs as well as Floating Car Data (FCD) messages as representative parameters of the information. For the simulation, the reference data D directly depends on the generated mobility pattern from the traffic simulator.

1.5.2. *Simulation set-up*

One particular aim of the simulation is the presentation of the features of our introduced cellular simulator VSimRTI_Cell. Thus, this simulator is part of the simulation set-up. In general, the set-up includes the following simulators for the different domains:

Traffic: the microscopic traffic simulator SUMO [KRA 12] simulates a realistic mobility pattern for the vehicles in the scenario;

Application: the VSimRTI internal simulator VSimRTI_App serves as a data generator for the communication messages and hosts the application logic for message reception and maintenance of a local dynamic map (LDM);

Ad hoc communication: the well-known network simulator OMNeT++ [VAR 08] simulates the IEEE 802.11p based communication stack and realistic radio propagation with fading and shadowing characteristics;

Cellular communication: the VSimRTI_Cell simulator, introduced in section 1.4, will simulate the transmission over the cellular network.

1.5.2.1. *Traffic simulation*

The scenario to be simulated is shown in Figure 1.7. It is located in an inner-city environment in Berlin (Germany). The scenario includes 30 reference vehicles overall to be equipped with the applications and the communication technologies. Only these reference vehicles are considered for the result evaluation. The vehicles are spawned into the simulation on ten different routes, although the routes could partially overlap. This means on each route at least three vehicles enable a sufficient grade of measurement coverage. The vehicles do not perform any reactions to the traffic situation like changing their route. The main intention is to drive their route and exchange information.

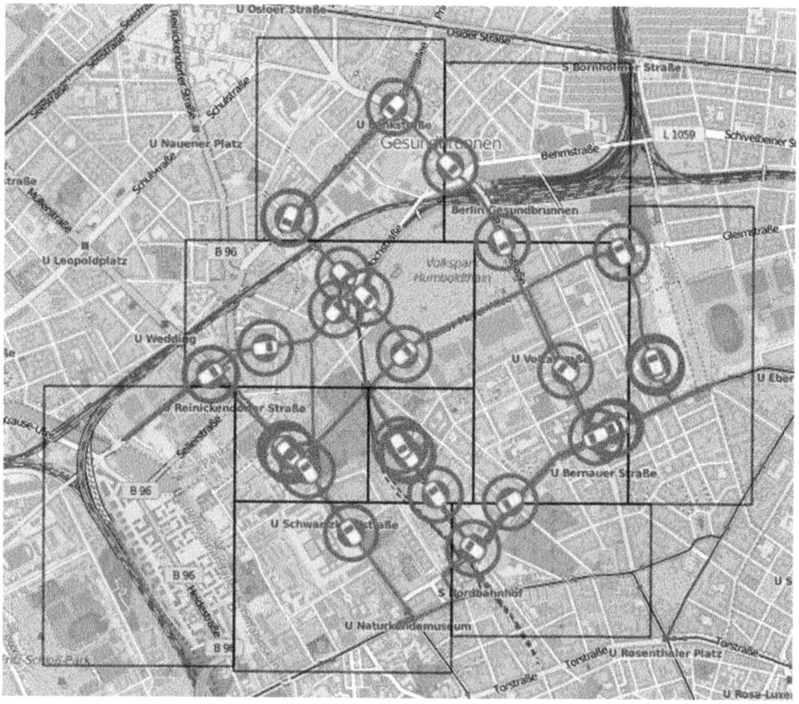

Figure 1.7. *Simulation scenario with routes of individual vehicles and cellular regions. For a color version of this figure, see www.iste.co.uk/hilt/transportation.zip*

1.5.2.2. *Application simulation*

The application logic is separated into three individual parts to be deployed on the vehicles and one application for a traffic efficiency server on the Internet. However, the simulated applications will not influence the traffic behavior with active route changing or similar actions:

VehicleMainModule implements the basic application facilities and should be equipped on the vehicle in every variation. It collects the sensor, location, speed and direction data to be included in the CAMs. Moreover, it maintains the LDM from sensor data as well as received messages from the ad hoc and cellular network. More specifically, the LDM implements a data matching of the information to a grid with geographic pixels;

VehicleAdhocModule uses the data from the VehicleMainModule and communicates it via IEEE 802.11p. It implements two different messages to be periodically disseminated. The CAMs only include the most recent local sensor data.

The FCD messages summarize the information in the LDM and map it to the central point of a geographic pixel before dissemination. Thus, it has two main parameters for the regular sending period of the CAMs and the FCD messages;

VehicleCellModule is the analogous component to the VehicleAdhocModule to communicate over the cellular network. This module supports an additional configuration for the local CAM destination area to be processed by the GEO in the cellular network. Moreover, it additionally sends CAMs per unicast to the Traffic Server. However, this application does not send FCD messages as they are managed centrally by the ServerModule;

ServerModule is the application on the server and maintains a central map with the same configuration as the LDM. It collects traffic information of the CAMs from the registered vehicles and periodically disseminates FCD messages back to the vehicles.

Table 1.1 outlines the specific configurations for the most important parameters of the individual application modules. Some parameters apply for multiple application modules.

Parameter	Application module	Value
LDM Grid Size	VehicleMainModule, ServerModule	20 × 20 pixels
LDM Pixel Side Length	VehicleMainModule, ServerModule	200 m
CAM Interval	VehicleAdhocModule, VehicleCellModule	100 ms
CAM Geo Radius	VehicleCellModule	695 m
CAM2Server Interval	VehicleCellModule	1 s
FCD Interval	VehicleAdhocModule, ServerModule	10 s

Table 1.1. *Simulation parameters for the application modules*

1.5.2.3. *Communication simulation*

The communication networks are simulated by OMNeT++ (ad hoc) and VSimRTI_Cell (cellular).

OMNeT++ uses the advanced communication models for the site-specific propagation, particularly shadowing characteristics [PRO 14c]. Moreover, OMNeT++ simulates the IEEE 802.11p based communication stack with the parameterization from Table 1.2. The given models for MAC and PHY layers respect all important aspects such as hidden terminals.

VSimRTI_Cell simulates the different cellular regions, shown as black rectangles in Figure 1.7. The region locations and expansions conform to data from OpenCellID (http://opencellid.org). All regions possess equal parameterizations for the communication properties. The configuration is presented in Table 1.2. It assumes an up-to-date HSPA network with capacity and delay properties to be in-line with recent measurements [SER 09, PRO 09, TEN 10]. The Traffic Server is located in a specific region with the properties of the overall network to simulate a well-connected Internet server.

IEEE 802.11p parameter	Value
Carrier Frequency	5.9 GHz
Bitrate	6 Mbit/s
TxPower	50 mW
RxSensitivity	–85 dBm
ThermalNoise	–94 dBm
AntennaGains	0 dBm
Cellular parameter	**Value**
Region UL Capacity	28.0 MBit/s
Region DL Capacity	42.2 MBit/s
Region DelayModel	GammaSpeedDelay
Region UL/DL minDelay	40 ms
Region UL/DL expDelay	150 ms
Network UL/DL Capacity	100 MBit/s
Network DelayModel	SimpleRandomDelay
Network UL/DL minDelay	10 ms
Network UL/DL maxDelay	30 ms
Network UL/DL delaySteps	3

Table 1.2. *Simulation parameters for the communication properties*

1.5.2.4. *Simulation variations*

For the subsequent simulation series, we investigate three different scenarios where all reference vehicles in the simulation are equipped with a variation of the application modules:

ad hoc VehicleMainModule + VehicleAdhocModule

cellular VehicleMainModule + VehicleCellModule

hybrid VehicleMainModule + VehicleAdhocModule + VehicleCellModule

The Internet-based traffic server is equipped in all scenarios with the ServerModule. However, in the ad hoc scenario, it never receives any messages.

1.5.3. *Simulation results*

In the following, we first analyze the safety capabilities of the different communication approaches with the help of the presented metric of the safe time ratio (STR). Afterwards, we evaluate the mean squared error (MSE) to measure the quality of the general information dissemination over a longer range in the whole scenario. Most traffic efficiency applications are usually based on such a dissemination principle.

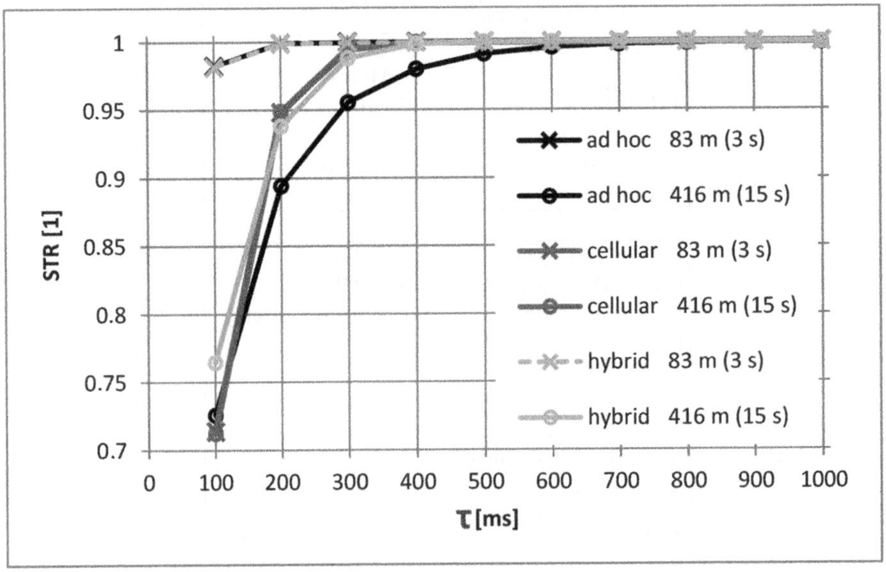

Figure 1.8. *Results for safety metric STR for different equipment settings and relevance areas*

1.5.3.1. *Safety metric*

The results for the STR are presented in Figure 1.8. The graph shows the STR's dependency on the real-time requirement τ. They include two variations: first, they show the communication technologies (ad hoc, black; cellular, dark gray; hybrid, light gray). Second, each access technology graph is presented with two different parameters for the relevance area time (t_R). In our evaluation, we define the relevance area according to the linear distance between the two vehicles. However, it could also incorporate further parameters as a converging trajectory, the same road or even lane etc. to limit the area to a more restricted set of relevant vehicles (e.g. eliminate vehicles in the opposite direction on a motorway). The linear distance, nonetheless, includes the most demanding properties. We selected a near-field relevance area of 83 m (line marker "x"), which addresses, according to Figure 1.6, use cases in the zone between Maneuver and Warning. For instance, the Intersection Collision Warning or the Electronic Brake Light Warning would be in this area. The second relevance area of 416 m (line marker "o") is in the middle of the Awareness zone from 1.6 and accounts for safety use cases with a slightly longer horizon, such as the Approaching Emergency Vehicle Warning.

The results for the ad hoc case in the near-field relevance area show that the STR already starts with a sufficiently high value of 98 % even for the most demanding τ of 100 ms. It quickly converges towards 100 % with a more relaxed τ. This is a result of the good communication properties of the direct IEEE 802.11p broadcasting with very short delays in the order of low ms and the low packet losses over short distances. The figures change for the medium field relevance area of 416 m, which should still be well within the limits of the communication range of our IEEE 802.11p configuration (with the parameters of transmission power, receiver sensitivity, etc.). However, the results reveal the known PHY Layer issues of increased packet loss due to fading, shadowing and also MAC layer coordination issues such as collisions due to the hidden terminal problem. Even in our moderate scenario, we could measure burst errors of spans longer than seven consecutive CAMs, resulting in the STR graph only converging toward 100 % at a τ of 700 ms. For higher relevance areas, the figures would turn out even more critical.

For the cellular case, both STR graphs show an equal trend, which is independent from the relevance distance. This reflects the expectation value of the underlying models of the regions with sufficient capacities to deliver all transmitted CAMs with the given delay distribution. We can see that there is a certain probability that messages are received out-of-order when particular messages, for example, take a longer way through the network with a higher latency. As the considered safety use cases mainly require the most recent updates of the CAMs only, older messages are dropped and neglected for the CCP and STR evaluation. This means, even when the data throughput of cellular networks is acceptable, the delay limits the performance of the use cases with real-time requirements less than 400 ms. This could be critical especially for use cases in the near-field relevance area of 83 m, where ad hoc communication shows its advantages of short latencies. If the future 5th generation of cellular networks can reduce latencies to the required scale, they could be a serious alternative for safety use cases.

For now, the hybrid approach to sending CAMs via ad hoc and cellular networks could be used as a migration path. The hybrid approach shows a similar trend to the cellular approach for the higher relevance distance of 416 m in supporting use cases with a τ of 400 ms fully with 100 %. It even starts at higher figures for the most demanding τ of 100 ms as the reception of short-delay ad hoc messages improves the performance. For the near-field relevance distance of 83 m, the ad hoc transmission appears to be dominating. Due to this, the hybrid approach delivers a more equal result compared to the ad hoc approach.

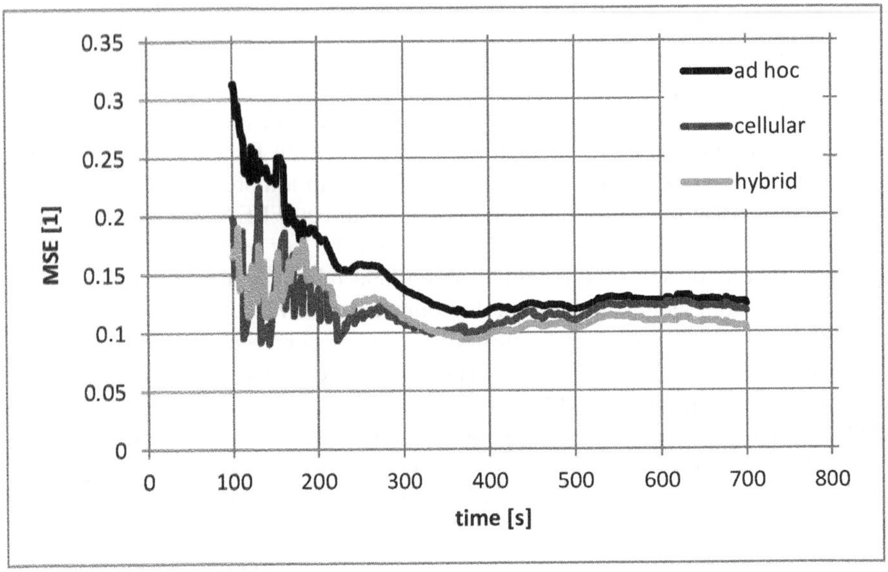

Figure 1.9. *Results for efficiency metric MSE for different equipment settings*

1.5.3.2. *Efficiency metric*

Figure 1.9 shows the development of the normalized MSE on the vehicles during the simulation time. It includes three graphs for the three different communication approaches (ad hoc, black; cellular, dark gray; hybrid, light gray). We cut away the very early and final phases of the simulation, when many vehicles still have to enter or, respectively, have already left the simulation. However, it is still worth examining the transition phases which would depict situations where the vehicles and thus the traffic information are not well distributed, but concentrated locally. Such situations may for instance appear temporarily in low traffic periods or in the early stage of system introduction when the penetration rate is generally low.

The trend of all graphs shows that the MSE generally decreases over the simulation time. It very slightly increases in the final phase when the first vehicles leave the simulation. We can see that the ad hoc approach, despite the short possible communication range, even reaches similar figures for a later simulation time compared to the other approaches. Our information handling algorithm which is based on the LDM actually implements a typical store-and-forward semantic. This approach collects information and carries it with the movement of the vehicle to later retransmit the summarized information. This is a very efficient method to increase the dissemination area for more delay-tolerant information. However, the black graph for ad hoc communication takes a longer time span to decrease as the vehicles have to drive for a certain amount of time to meet and exchange the information they have collected on their way. In comparison, the graphs for the cellular and hybrid approaches already start at fairly lower MSE values in the beginning of the simulation. This is due to the fact that the Traffic Server can quickly mirror the perceived traffic information back to the equipped vehicles. In the later simulation time, the hybrid approach slightly outperforms the cellular approach.

In summary, it could be stated that the cellular approach in this time period already delivers sufficient results for information dissemination. The hybrid approach, with additional messages over ad hoc communication, may still improve the redundancy. Nonetheless, our presented information handling application is still very simple and could be still improved with more advanced techniques for data aggregation, e.g. from the field of machine learning. However, this was out of scope of the presented evaluation.

1.6. Conclusion

Ad hoc networks based on IEEE 802.11p enable a decentralized information exchange among vehicles, and among vehicles and infrastructure units. Since the limited communication range and the lack of deterministic quality of service are a challenge for the scalability of ad hoc networks, some new approaches try to overcome these drawbacks by using cellular networks for the information exchange among vehicles. However, although cellular networks enable a nearly unlimited communication range, the architecture of these networks can involve a delay in information transmission which might violate the strong requirements of many safety applications. To reduce the drawbacks of both networks types, an intelligent combination of vehicular ad hoc networks and cellular networks could help. However, detailed analyses are needed to evaluate in which cases pure ad hoc networks, pure cellular networks or a combination of both would be the best. To give the research community a powerful tool for these evaluations, we have developed the novel cellular communication simulator VSimRTI_Cell. This lightweight tool models a level of abstraction of cellular networks and allows the simulation of large scenarios. Due to the coupling of VSimRTI_Cell to the existing simulation framework VSimRTI, this extended framework is predestined for the modeling of ad hoc networks and cellular networks. Our simulation study, presented in this work, gives an example to show how the research in this area can be addressed.

1.7. Bibliography

[CAR 07] CAR 2 CAR COMMUNICATION CONSORTIUM, C2C-CC Manifesto – Overview of the C2C-CC System, num. Ver 1.1, August 2007.

[ELB 06] ELBATT T., GOEL S.K., HOLLAND G. *et al.*, "Cooperative collision warning using dedicated short range wireless communications", *Proceedings of the 3rd International Workshop on Vehicular Ad Hoc Networks*, ACM, pp. 1–9, 2006.

[ETS 09] ETSI, ETSI TR 102 638: Intelligent Transport Systems (ITS); Vehicular Communications; Basic Set of Applications; Definitions, Technical Report num. Ver 1.1.1, European Telecommunications Standards Institute, June 2009.

[ETS 10] ETSI, ETSI TS 102 637-1: Intelligent Transport Systems (ITS); Vehicular Communications; Basic Set of Applications; Part 1: Functional Requirements, Technical Specification num. Ver 1.1.1, European Telecommunications Standards Institute, September 2010.

[ETS 13] ETSI, ETSI TR 101 539-3: Intelligent Transport Systems (ITS); V2X Applications; Part 3: Longitudinal Collision Risk Warning (LCRW) application requirements specification, Technical Specification num. Ver 1.1.1, European Telecommunications Standards Institute, November 2013.

[ETS 14a] ETSI, ETSI EN 302 637-2: Intelligent Transport Systems (ITS); Vehicular Communications; Basic Set of Applications; Part 2: Specification of Cooperative Awareness Basic Service, European Standard num. Ver 1.3.2, European Telecommunications Standards Institute, November 2014.

[ETS 14b] ETSI, ETSI EN 302 637-3: Intelligent Transport Systems (ITS); Vehicular Communications; Basic Set of Applications; Part 3: Specifications of Decentralized Environmental Notification Basic Service, European Standard num. Ver 1.2.2, European Telecommunications Standards Institute, November 2014.

[ETS 14c] ETSI, ETSI EN 302 895: Intelligent Transport Systems (ITS); Vehicular Communications; Basic Set of Applications; Local Dynamic Map (LDM), European Standard num. Ver 1.1.1, European Telecommunications Standards Institute, September 2014.

[GOE 14] GOEBEL N., KOEGEL M., MAUVE M. *et al.*, "Trace-based simulation of C2X-communication using cellular networks", *11th Annual Conference on Wireless On-demand Network Systems and Services (WONS)*, IEEE, pp. 108–115, 2014.

[HEN 08] HENDERSON T.R., LACAGE M., RILEY G.F. *et al.*, "Network simulations with the ns-3 simulator", *SIGCOMM Demonstration*, vol. 15, p. 17, 2008.

[IEE 10] IEEE, IEEE Std 1516-2010 (Revision of IEEE Std 1516-2000): IEEE Standard for Modeling and Simulation (M&S) High Level Architecture (HLA) – Framework and Rules, Std, IEEE Computer Society, August 2010.

[IKU 10] IKUNO J.C., WRULICH M., RUPP M., "System level simulation of LTE networks", *IEEE 71st Vehicular Technology Conference*, IEEE, pp. 1–5, 2010.

[KLO 12] KLOIBER B., GARCIA C., HÄRRI J. *et al.*, "Update delay: a new information-centric metric for a combined communication and application level reliability evaluation of cam based safety applications", *ITS World Congress*, 2012.

[KRA 12] KRAJZEWICZ D., ERDMANN J., BEHRISCH M. *et al.*, "Recent development and applications of SUMO–simulation of urban mobility", *International Journal on Advances in Systems and Measurements*, vol. 5, no. 3 and 4, pp. 128–138, 2012.

[NAU 09] NAUMANN N., SCHÜNEMANN B., RADUSCH I. *et al.*, "Improving V2X simulation performance with optimistic synchronization", *IEEE Asia-Pacific Services Computing Conference*, pp. 52–57, December 2009.

[PIR 11] PIRO G., GRIECO L.A., BOGGIA G. *et al.*, "Simulating LTE cellular systems: an open-source framework", *IEEE Transactions on Vehicular Technology*, vol. 60, no. 2, pp. 498–513, 2011.

[PRO 09] PROKKOLA J., PERÄLÄ P.H., HANSKI M. *et al.*, "3G/HSPA performance in live networks from the end user perspective", *IEEE International Conference on Communications*, pp. 1–6, 2009.

[PRO 11] PROTZMANN R., SCHÜNEMANN B., RADUSCH I., "The influences of communication models on the simulated effectiveness of V2X applications", *Communications Magazine, IEEE*, vol. 49, no. 11, pp. 149–155, 2011.

[PRO 14a] PROTZMANN R., MASSOW K., RADUSCH I., "An evaluation environment and methodology for automotive media streaming applications", *Eighth International Conference on Innovative Mobile and Internet Services in Ubiquitous Computing (IMIS)*, IEEE, pp. 297–304, 2014.

[PRO 14b] PROTZMANN R., MASSOW K., RADUSCH I., "On performance estimation of prefetching algorithms for streaming content in automotive environments", *11th Annual Conference on Wireless on-demand Network Systems and Services (WONS)*, IEEE, p. 147, 2014.

[PRO 14c] PROTZMANN R., SCHÜNEMANN B., RADUSCH I., "On site-specific propagation models for the evaluation of V2X applications", *7th International Workshop on Communication Technologies for Vehicles (Nets4Cars-Fall)*, IEEE, pp. 35–39, 2014.

[PRO 14d] PROTZMANN R., SCHÜNEMANN B., RADUSCH I., "A sensitive metric for the assessment of vehicular communication applications", *IEEE 28th International Conference on Advanced Information Networking and Applications (AINA)*, IEEE, pp. 697–703, 2014.

[QUE 08] QUECK T., SCHÜNEMANN B., RADUSCH I. *et al.*, "Realistic simulation of V2X communication scenarios", *APSCC '08: Proceedings of the 2008 IEEE Asia-Pacific Services Computing Conference*, IEEE Computer Society, Washington, pp. 1623–1627, 2008.

[SCH 11] SCHÜNEMANN B., "V2X simulation runtime infrastructure VSimRTI: an assessment tool to design smart traffic management systems", *Computer Networks*, vol. 55, pp. 3189–3198, Elsevier North-Holland Inc., 2011.

[SEG 14] SEGATA M., BLOESSL B., JOERER S. *et al.*, "Towards inter-vehicle communication strategies for platooning support", *7th International Workshop on Communication Technologies for Vehicles (Nets4Cars-Fall)*, IEEE, pp. 1–6, 2014.

[SER 09] SERRANO C., GARRIGA B., VELASCO J. *et al.*, "Latency in broad-band mobile networks", *IEEE 69th Vehicular Technology Conference*, VTC Spring 2009, IEEE, pp. 1–7, 2009.

[TEN 10] TENORIO S., EXADAKTYLOS K., MCWILLIAMS B. *et al.*, "Mobile broadband field network performance with HSPA+", *Wireless Conference (EW)*, IEEE, pp. 269–273, 2010.

[VAR 08] VARGA A., HORNIG R., "An overview of the OMNeT++ simulation environment", *Proceedings of the 1st International Conference on Simulation Tools and Techniques for Communications, Networks and Systems & Workshops*, ICST (Institute for Computer Sciences, Social-Informatics and Telecommunications Engineering), p. 60, 2008.

[VIR 14] VIRDIS A., STEA G., NARDINI G., "SimuLTE – a modular system-level simulator for LTE/LTE-A networks based on OMNeT++", *International Conference on Simulation and Modeling Methodologies, Technologies and Applications (SIMULTECH)*, pp. 59–70, August 2014.

[WED 09] WEDEL J.W., SCHÜNEMANN B., RADUSCH I., "V2X-based traffic congestion recognition and avoidance", *International Symposium on Parallel Architectures, Algorithms, and Networks*, IEEE Computer Society, pp. 637–641, 2009.

2

Near-field Wireless Communications and their Role in Next Generation Transport Infrastructures: an Overview of Modelling Techniques

The development of the smart city (SC) paradigm relies on the need for more interconnected public Intelligent Transportation Systems (ITSs). In fact, nowadays, smart cyber physical systems in the transportation domain are expected to play an important role in the ambition to develop passenger-centric services. Like other utilities, transport infrastructures are slowly moving forward to more intelligent, connected, user-centric and collaborative systems. This movement is partly supported by the increasing availability of low-cost smart objects with wireless interconnection capabilities and wireless indoor positioning systems. A significant revolution has been envisaged in the transportation domain by the introduction of these low-cost elements with their wireless interconnection capabilities mostly in the near-field environment. In fact, cars, trains, buses, bicycles and road infrastructures are becoming increasingly equipped with sensors, RFID tags and NFC devices, sending critical information to the traffic control centers to better route traffic and to provide users with real-time relevant transportation information.

New research, engineering methods, tools and simulation studies for this cyber-physical—and near-field—scenario in the transportation domain have to be developed. This chapter is outlined in the context of modeling techniques to explore this challenging Big Data or Internet of Things (IoT) scenario in the ITS domain. Simulation modeling is a necessary and crucial step in the design, development, test

Chapter written by Christian PINEDO, Marina AGUADO, Lara RODRIGUEZ, Iñigo ADIN, Jaizki MENDIZABAL and Guillermo BISTUÉ.

and performance evaluation of any communication network implementation or strategy. To provide a global performance study of the IoT environment in a specific transportation scenario, a system level simulation approach is necessary. Traditionally, system level simulators receive results from link level simulation studies as an input. However, well-known discrete event system level network simulators have so far not included near-field wireless communication behavior despite the fact that these low-cost smart devices are widespread and are relevant actors who impact and stress the communication infrastructures.

Our goal, in this chapter, is to provide guidelines on modeling these smart low-cost near-field wireless objects and on how to integrate their behavior in traditional network Discrete Event Simulation (DES) tools. The ultimate aim is to provide an insight into the available tools in order to study—with an overall perspective—their behavior and their impact on the access and core communication infrastructures of ITS.

This chapter is structured as follows. Section 2.1 provides an overview of near-field wireless technologies, introducing the near-field and far-field concepts together with a taxonomy of the near-field wireless technologies which can be found in the transportation domain. Section 2.2 introduces the two main existent techniques for the characterization of the near-field communication link. This link level characterization traditionally serves as an input to system level simulation techniques based on DES tools. Section 2.3 presents the state of the art of these DES frameworks and their different approaches to near-field modeling for performance evaluation purposes. The last section covers the main conclusions and further research opportunities.

2.1. Near-field wireless technologies

This section deals with the definition of near-field compared to far-field in communications with the aim of providing the reader with a better technical understanding of these concepts. Then, the most relevant near-field wireless technologies that are being used today for a number of transport applications are detailed.

2.1.1. *Near-field versus far-field*

Near-field and far-field concepts are related to the generation of an electromagnetic field in the area surrounding an antenna, where an alternate current flowing through a conductor loop mainly generates a magnetic field (H), and an alternate current flowing through a conductor dipole mainly generates an electric field (E). As these fields (H or E) propagate, an electromagnetic field is created (a field composed of electric and magnetic fields). The interaction between these fields creates an electromagnetic wave able to travel into space.

Depending on the distance from the source, the field that surrounds an antenna can be broken up into two segments: near-field and far-field. Typically, near-field is defined as the field around the antenna up to $\lambda/2\pi$ away, where λ is the wavelength. After this point, the electromagnetic wave begins to separate from the antenna and therefore, the ability to interact by inductive or capacitive coupling is lost, reaching the far-field zone after a transition zone. Figure 2.1 shows the different zones and their names.

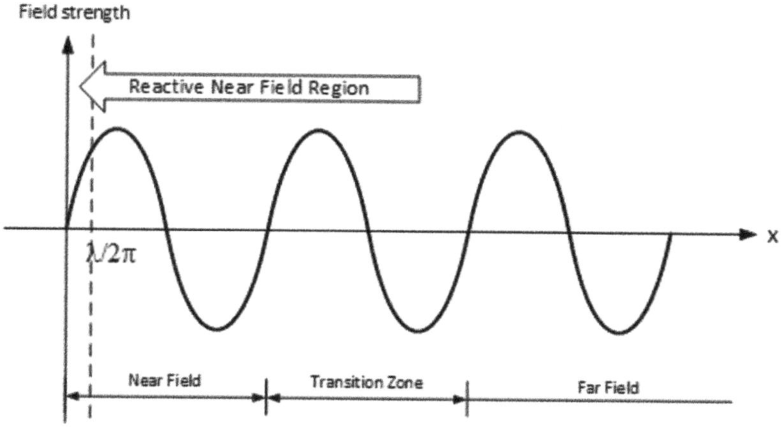

Figure 2.1. *Near-field versus far-field*

Near-field and far-field have different energies so they typically require a corresponding antenna type because the near-field applications primarily employ the magnetic field, while the far-field has both electric and magnetic components. When moving from near-field to far-field, the wave impedance varies, being constant at a value of 377Ω in the far-field region as shown in Figure 2.2, together with the different zones.

The closest region to the antenna is known as near-field reactive. In this region, the most important characteristic is the presence of a dominant magnitude. Depending on the physical characteristics of the antenna, one field (electric or magnetic) will prevail over the other. More specifically, in the case of two rectangular loops, the coupling may be characterized taking just the magnetic field into account.

The region beyond two wavelengths is called the far-field, and in this case the electric (E) and magnetic (H) fields support and regenerate one another as their strength decreases inversely as the square of the distance. Table 2.1 shows the distance of EM field separation ($\lambda/2\pi$) for several frequencies associated with communication systems.

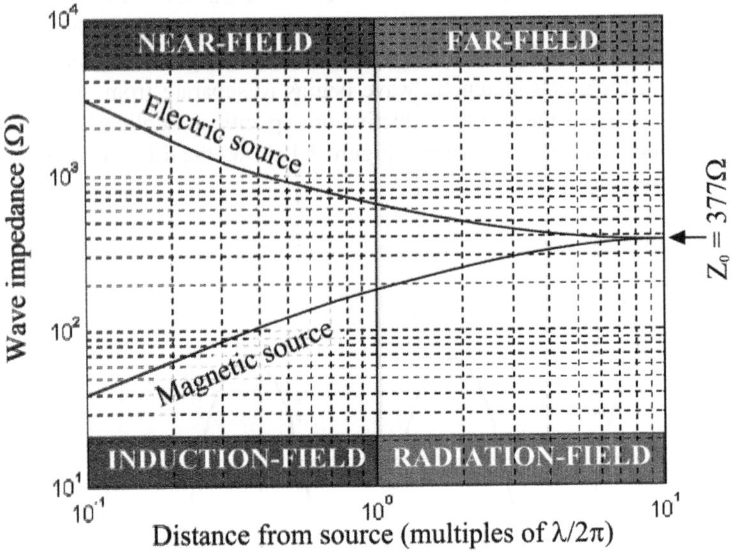

Figure 2.2. *Wave impedance (Dannex HF-equipment Sweden, http://www.dannex.se/theory/3.html)*

Frequency	$\lambda/2\pi$
4 MHz	1193 cm
27 MHz	177 cm
433 MHz	11 cm
868 MHz	5.5 cm
915 MHz	5.2 cm
2.45 GHz	1.9 cm
5.8 GHz	0.8 cm

Table 2.1. *Value of $\lambda/2\pi$ (cm) for different wavelengths*

Working in near-field or in far-field differs as summarized as follows:

– the coupling between two antennas in the near-field (reactive) is similar to the AC transformer, whereas the far-field coupling is commonly described as RF communication;

– in near-field communications, the received signal mainly depends on the characteristics of the source. In far-field links, the most relevant parameter is the communication channel;

– inductive coupling is mainly determined by the relative distances of transmitter and receiver parts, whereas radiation is usually more closely related to differences in propagation time and/or path.

As mentioned, wireless communications (via the antenna) occur using a process known as electromagnetic coupling. There are two types of coupling:

– inductive: a near-field antenna uses inductive coupling which means that it uses a magnetic field (H) or an electric field (E). A magnetic (H) or electric (E) field is created in the near-field region that allows the antenna to transmit the signal;

– capacitive: a far-field antenna uses capacitive coupling (or propagation coupling). Capacitive coupling occurs when the signal sent by the antenna propagates and an electromagnetic signal is available.

In contrast to far-field antennas which transmit a propagating electromagnetic field, a near-field antenna generates a local magnetic (H) or electric (E) field suitable for short read-range applications. Near-field antennas are not very commonly used and therefore there are limited options available in the market. On the other hand, far-field antennas present a wide variety of shapes and sizes offering a larger coverage than the near-field antennas. Many options are available regarding far-field antennas such as linear or circular polarization, varying gain, indoor or outdoor use, and multi-band; many wireless technologies are based on far-field communications, such as Zigbee and Bluetooth.

2.1.2. *Near-field-based technologies in transport*

Specific near-field wireless communication technologies have been used for a long time in transport systems such as trains, undergrounds and trams.

In this sense, balises are very common devices present in many Automatic Train Protection (ATP) systems. Balises' role is to increase safety and avoid collisions or other kinds of accidents like derailments due to high speed. Balises are powerless devices located on the track and they are telepowered by the train when the train passes by. Once the balise is active, it transmits information—an inductive communication—to the Balise Transmission Module (BTM) of the train to warn the driver or even to stop the train if the driver is not acting as expected. Balises are not only a fundamental part of many national ATP systems such as the Spanish ASFA or the German PZB, but they are also part of the most recent and modern ATP systems such as the European Train Control System (ETCS)—where the balise is called *Eurobalise* [UNI 12]—or Communications-Based Train Control (CBTC).

Euroloop [UNI 08] is another example of inductive technology used in ETCS. It is quite similar to a balise. In fact, it is an extension of the Eurobalise over a particular distance to be able to continuously transmit data to the Loop Transmission Module (LTM) of the vehicle over cables emitting electromagnetic waves in a similar way to other national systems such as the German LZB or Thales' Euroloop.

Apart from these traditional railway technologies, transportation systems are also embracing the SC and IoT paradigms. In fact, in [GUB 13], the authors introduce the concept of *smart transportation* and *smart logistics* as an area of applicability of IoT concepts. Some of the services pointed out in [ZAN 14] are related to these domains such as *traffic management* and *smart parking*. Furthermore, in [ATZ 10], the authors add the following services and applications: *assisted driving*, *mobile ticketing*, *monitoring environmental parameters* and *augmented maps*.

The number of communication technologies that can be used in the SC and IoT domains is huge and it depends on the physical characteristics of the devices and the specific use case [ZAN 14]. Some commonly used communication technologies are wired (e.g. Ethernet, Fiber Optic), whereas others are wireless (e.g. WiFi, UMTS, LTE). Only two of those wireless technologies can be considered inductive: RFID and NFC.

In fact, nowadays RFID and NFC are two of the most common near-field wireless technologies and, thus, they are detailed in the following subsections.

2.1.2.1. *RFID*

Radio Frequency IDentification (RFID) [WAN 06] is an automatic wireless data collection technology with a long history [LAN 05]. This technology is usually employed to identify items by means of radio waves. The basic composition of an RFID system covers a tag and a reader. The reader sends an interrogating signal to the tag, and the tag responds with its unique information. RFID tags are classified into Active, Semi-Passive or Passive:

– Active RFID tags contain a battery and therefore they rely on their own power source. As a result, the active tag can be read by signals up to 100 meters. This long read range makes active RFID tags ideal for many industries where asset location and other improvements in logistics are important. Active tags may be either read-only or read/write, thus allowing data modification by the reader. Other benefits such as data storage and faster data rates make this kind of tag very appropriate for electronic toll collection. There is an endless variety of tags. This kind of tag is the most expensive one;

– Semi-passive RFID tags are similar to passive tags in using the reader signal to provoke a response from the tag, and similar to active tags in containing a battery to power all the electronics of the tag itself. Usually, the semi-passive tag presents a longer operating life in terms of power supply compared to active tags, but on the other hand these tags have some of the limitations of the passive tag in terms of slow read speeds and short read distances. The price of semi-passive tags is lower than the active tags and higher than the passive tags;

– Passive RFID tags do not contain any power source. Instead, the energy employed to power these tags is the energy of the electromagnetic signal sent by the RFID reader. Therefore, passive tags depend on the reader's RF signal to respond. Usually, passive RFID tags have a read range from near contact and up to 25 meters. Currently, the most employed type of RFID tag is the passive one. In general, their design is simpler and does not contain a battery. The tag may be used in many applications thanks to the different forms it can take, ranging from identification cards for public transportation to tags embedded in license plates for car identification. This type of tag is the cheapest one. In a passive RFID system, the reader transmits a modulated RF signal to the tag consisting of an antenna and an integrated circuit chip. The chip receives power from the antenna and responds by varying its input impedance and thus modulating the backscattered signal. There were functional passive RFID systems already being reported in the early 1970s [KOE 75]. Since then, RFID has advanced [FIN 04, KAR 03, GLI 04, DEV 05] and has experienced tremendous growth.

RFID tags primarily operate at three frequency ranges. These frequency ranges also set the type of communication that can be employed:

– Low Frequency (LF) 125–134 kHz;

– High Frequency (HF) 13.56 MHz;

– Ultra-High Frequency (UHF) 856–960 MHz.

Low-frequency (LF, 125–134 KHz) and high-frequency (HF, 13.56 MHz) [EPC 13] RFID systems are short-range systems based on inductive coupling between the reader and the tag antennas through a magnetic field (near-field).

Ultra-high frequency (UHF, 860–960 MHz) [EPC 13] and microwave (2.4 GHz and 5.8 GHz) RFID systems are long-range systems which use electromagnetic waves propagating between reader and tag antennas (far-field). EPC provides a specification for this kind of RFID.

There are many RFID standards depending on the application RFID is intended to be used for. DIN, ISO and VDE are some normalization bodies offering these standards.

2.1.2.2. *NFC*

Near-Field Communication (NFC) is a specific subset of High-Frequency (HF) RFID. NFC allows for the secure exchange of data. Moreover, an NFC device acts as an NFC reader and NFC tag. Therefore, NFC devices are able to communicate peer-to-peer.

NFC devices operate at the same 13.56 MHz frequency as HF RFID readers and tags. The standards and protocols of the NFC format deal with the use of RFID in proximity cards and are based on RFID standards [ISO 16, ISO 13].

As a finely honed version of HF RFID, near-field communication devices have taken advantage of the short read-range limitations of its radio frequency. Because NFC devices must be in close proximity to each other, usually no more than a few centimeters, it has become a popular choice for secure communication between consumer devices such as smartphones.

Peer-to-peer communication is a feature that sets NFC apart from typical RFID devices. An NFC device is able to act both as a reader and as a tag. This unique ability has made NFC a popular choice for contactless payment, a key driver in the decision by influential players in the mobile industry to include NFC in newer smartphones. Also, NFC smartphones pass along information from one smartphone to the other by tapping the two devices together, which turns sharing data such as contact information or photographs into a simple task. Recently, you may have seen advertising campaigns that use smart posters to pass information to the consumers.

Also, NFC devices can read passive NFC tags, and some NFC devices are able to read passive HF RFID tags that are compliant with ISO 15693. The data on these tags can contain commands for the device such as opening a specific mobile application. You may start seeing HF RFID tags and NFC tags more frequently in advertisements, posters and signs as it is an efficient method to pass along information to consumers.

In effect, NFC builds upon the standards of HF RFID and turns the limitations of its operating frequency into a unique feature of near-field communication.

2.2. Characterization of near-field communications

Currently two different approaches can be distinguished regarding the characterization of near-field communications, namely theoretical calculations based on electrical models and electromagnetic simulators. Although the latter also rely on the electrical models, the way of working with both of them differs. For the electrical models, mathematical equations are derived, and therefore a huge mathematical background is required. On the other hand, simulators allow us to graphically model the elements to be characterized and the computation is done by the simulator itself, in the time domain or in the frequency domain.

As a result of the characterization, a transfer function of the near-field communication is obtained. This transfer function can be employed in the DES framework as the model of the link level near-field zone between the communication devices.

2.2.1. *Electrical models*

Figure 2.3 shows the basic electric model used for inductive loops [DOB 12]. It represents a simplified model of a transformer, incorporating the particular elements found in this field.

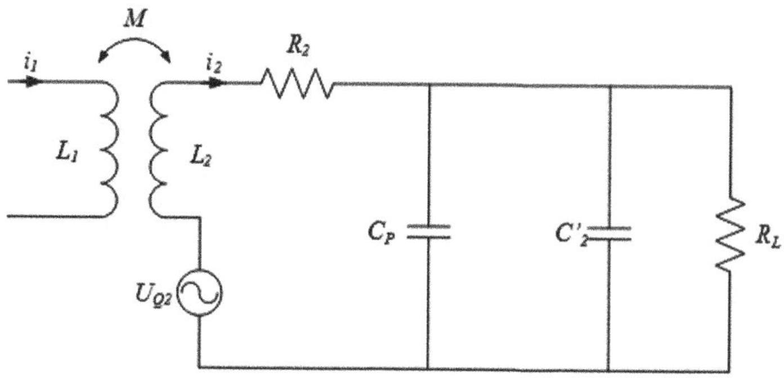

Figure 2.3. *Simplified equivalent circuit for inductive loops*

In this figure, L_1 and L_2 are the inductances of each coil, U_{Q2} is the voltage induced in the coil 2, R_1 and R_2 are the ohmic losses of each coil, C_p is the parasitic capacitance of loop 2 and C_2 is the tuning capacitor. C_2 is used to provide the appropriated resonant frequency.

One of the key points of the model is the analysis of R_L. In [CHA 10] the R_L of an UHF RFID passive tag is presented, showing the differences in near- and far-fields. In the case of many commercially available devices, when the tag is located in the near-field the received antenna power increases, therefore the system includes some kind of nonlinear control circuit to avoid potential damages. Another interesting model is shown in [JAN 11], including a simple description of each building block.

In any case, the performance of the complete system is conditioned by the ability of the emitting antenna to induce voltage into the passive loop. The next point presents the analysis of one of the most frequent cases used in RFID communications.

2.2.2. *Analysis of the mutual inductance of a squared inductive coupling*

Taking into account the Biot-Savart law [2.1], the magnetic field created by any segment of the loop at any point of the space can be calculated as follows:

$$d\vec{B} = \frac{\mu_0 I}{4\pi} \frac{(\vec{dl} \Lambda \vec{u}_r)}{r^2} \qquad [2.1]$$

where $\vec{dl}$ represents a differential element of the conductor carrying a current I, $\vec{u}_r$ is the unit vector in the direction of the straight line between the differential element and the point of interest, and r is the distance between these items.

[2.1] allows us to compute the magnetic field created by a rectangular loop at any point of the space. In the case of rectangular loops, it is very important to know the z-component of the field.

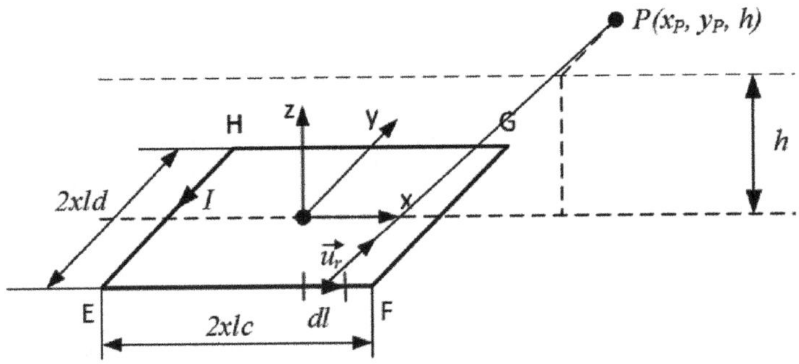

Figure 2.4. *Magnetic field created by a rectangular loop at point P*

In order to calculate the magnetic field created by the EF side of the loop shown in Figure 2.4, we must introduce the expressions [2.2], [2.3] and [2.4] in [2.1]:

$$\vec{dl} = (dx \; 0 \; 0) \quad [2.2]$$

$$\vec{u}_r = \left(\frac{x - x_p}{r} \; \frac{-l_d - y_p}{r} \; \frac{0 - h}{r}\right) \quad [2.3]$$

$$r = \sqrt{(x - x_p)^2 + (y_p - ld)^2 + h^2} \quad [2.4]$$

The differential field created by $\vec{dl}$ is given by:

$$\vec{dB}_{EFz} = \frac{\mu_0 I}{4\pi r^2} \begin{vmatrix} \vec{i} & \vec{j} & \vec{k} \\ dx & 0 & 0 \\ \frac{x - x_p}{r} & \frac{-l_d - y_p}{r} & \frac{0 - h}{r} 0 \end{vmatrix} = -\frac{\mu_0 I}{4\pi r^3}(ld + y_p) dx \vec{k} \quad [2.5]$$

The integral of expression [2.5] over the EF segment gives the total field created:

$$B_{EFz}(x_p, y_p) = \int_{-lc}^{lc} dB_{EFz} = \int_{-lc}^{lc} \frac{\mu_0 I}{4\pi} \frac{(ld + y_p)}{[(x - x_p)^2 + (y_p + ld)^2 + h^2]^{3/2}} dx \quad [2.6]$$

The field created by the rest of the elements of the loop can be calculated in the same way. Finally, expression [2.7] will enable the calculation of the vertical component of the magnetic field at any point.

$$B_{EFGHz}(x_p, y_p) = B_{EFz}(x_p, y_p) + B_{FGz}(x_p, y_p) + B_{GHz}(x_p, y_p) + B_{HEz}(x_p, y_p) \quad [2.7]$$

In order to characterize the interaction between two rectangular loops, the situation shown in Figure 2.5 can be considered. It is considered a distance Δx between the centers of both loops, and a perfect alignment between their longitudinal axes.

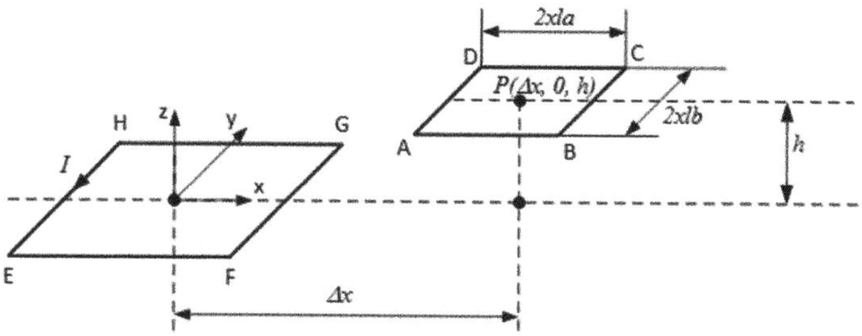

Figure 2.5. *Inductive coupling of two x-axis aligned rectangular coils*

In order to calculate the voltage induced in loop ABCD, we must express the magnetic flux created by loop EFGH through loop ABCD as a function of time. We can achieve this goal combining expressions [2.8].

$$\Phi_{ABCD} = \int_{y_A}^{y_D} \int_{x_A}^{x_B} B_{EFGHz}(x_p, y_p) dx_p dy_p$$

where

$$x_A = \Delta x - la$$
$$x_B = \Delta x + la \quad [2.8]$$
$$y_A = -lb$$
$$y_D = lb$$

The magnetic flux will be given by [2.9]:

$$\Phi_{ABCD}(\Delta x) = \int_{y_A}^{y_D} \int_{\Delta x - la}^{\Delta x + la} B_{EFGHz}(x_p, y_p) dx_p dy_p \quad [2.9]$$

The final step is to express the time-variant magnitudes of [2.9] as a function of time. In the particular case of loops for railway applications, there are two sources of time variations:

– the current of loop EFGH, usually given by a sine function;

– the center distance Δx, which may be expressed as a function of train speed.

In any case, operator [2.10] will give the final expression:

$$v = -\frac{d\Phi_{ABCD}(t)}{dt} \qquad [2.10]$$

2.2.3. *Computer-aided electromagnetic calculation*

The most powerful tool available today for electromagnetic analysis is the computer simulator. There are several commercial packages capable of performing accurate 3D simulations, taking into account different boundary conditions. These platforms can replicate virtually any physical system. However, there are two main drawbacks: on the one hand, the use of computational resources may be high even for simple models, and on the other hand, it may be difficult to relate the performance obtained with the basic design parameters.

It is also possible to combine the computer's computation capacity with the equations mentioned in the previous section. For example, BTM antennas are defined as rectangular loops, and therefore, the analysis of this particular system may be carried out by applying the basic electromagnetic equations to the system as shown in section 2.2.2. With this approach a greater insight into the physical problem is obtained, and as a result, the conclusions obtained can be used to optimize the system with quantitative and qualitative rules. In order to simplify the expressions, a symbolic mathematical solver can be used. This methodology requires less computational resources than the conventional electromagnetic simulator but it shows some constraints due to the simplifications required in the mathematical analysis.

Currently, electromagnetic simulation software is a tool that supports communication designers, obtaining accurate predictions for more complex structures than two-faced rectangular loops. There are a number of different electromagnetic analysis programs that differ from each other in a number of different underlying technologies. Each simulation technology offers particular benefits, and therefore solving a specific problem type requires the use of one particular type of electromagnetic simulator that best suits the problem.

Computer-Aided Engineering (CAE) software has only been used for around 25 years, although currently it is one of the key parts of the design process. Nowadays, efficient and powerful personal computers are capable of running highly computationally demanding CAE programs in a reasonable time. CAE tool developers have taken advantage of this computer performance improvement which has resulted in the availability of unprecedented levels of simulation capability. This is a significant advantage in the field of electromagnetic simulators since the problem sizes associated with solving Maxwell's equations can be quite large. Nevertheless, CAE simulation tools' performance constraints are generally not the speed of the simulation engine, but the accuracy or availability of the models employed for the simulation. Usually, designs can be classified into active devices represented by nonlinear models or passive devices that are represented by linear models. But since even passive components such as cables and connectors exhibit nonlinear behavior, complex models are often needed for them. Moreover, passive components can be classified into discrete or lumped components (resistors, capacitors, and inductors) and distributed components, such as those formed of microstrip transmission lines [SWA 03].

Electromagnetic simulators solve different circuit problems based on Maxwell's equations. Currently, most of the electromagnetic simulators rely on three key technologies: Method of Moment (MoM), Finite Element Method (FEM) and Finite Difference Time Domain (FDTD) methods. These simulation methods tend to use a similar approach to solve the problems:

– first, a physical model is created. This usually consists of layout geometries, material properties, etc.;

– then, the simulator is set up with the boundary conditions, the extent of the simulation and the assignment of ports and other specific simulation options;

– once the model is defined and the simulator is set up, the simulation is performed. The simulation involves the use of mesh cells to transform the physical model into discrete elements. The simulator makes us of local functions to approximate the field/current across the mesh cells;

– finally, the local function coefficients are adjusted until the boundary conditions of the simulation are fulfilled. Design information such as S-parameters, field level and/or radiation patterns can be calculated during post-processing.

This process is similar for simulators based on MoM, FEM and FDTD [HES 09]. However, the differences among them make each one best suited for particular applications:

MoM: this technique only requires that the metal interconnects in the structure to be simulated are meshed. Therefore, simulations are speed up compared to the other technologies because a “planar” MoM mesh is simpler and smaller than the equivalent “3D volume” mesh required for an FEM or FDTD simulation. MoM algorithms solve Maxwell’s equations implicitly by solving a matrix;

FEM: this simulation method is a true 3D field solver that allows arbitrarily shaped 3D structures to be analyzed. The advantage over MoM simulation is that it can be used for any type of 3D structure and is not confined to a layered stack up. FEM simulation requires that objects being simulated are placed into a truncated space. This volume of the simulation domain is converted into discrete elements, usually tetrahedral mesh cells with a denser mesh being created around the geometric model being simulated. FEM algorithms solve Maxwell’s equations implicitly by solving a matrix;

FDTD: this simulation method is a true 3D field solver which can analyze arbitrary shaped 3D structures like FEM. FDTD algorithms solve Maxwell’s equations in a fully explicit way. FDTD employs a time-stepping algorithm that updates the field values across the mesh cell time-step by time-step, thereby explicitly following the EM waves as they propagate through the structure modeled.

In order to select the most suitable EM simulators based on MoM, FEM and FDTD analysis methods for a given application, the geometry of the design and the circuit response type are the first parameters to consider:

– MoM-based simulators offer the most efficient simulation method for truly planar structures. For that reason, an MoM-based simulator would be recommended for analysis of on-chip passive elements and components on a PCB and planar antennas. However, it is not the best method for communication between two antennas or between a passive tag and a reader as in the case of RFID. Either FEM- or FDTD-based EM simulators are usually more appropriate for true 3D structures [HES 09];

– both MoM and FEM methods solve natively in the frequency domain, which makes them more appropriate than FDTD for the analysis of circuits with a high quality factor (high Q), such as filters, cavities, resonators and oscillators. In contrast, the FDTD method solves natively in the time domain, making them useful for connector interfaces and transitions.

2.3. Discrete event simulators

Wireless communications have attracted considerable interest in the research community, and many wireless networks are evaluated using Discrete Event Simulation (DES) tools. This section examines different simulation tools and makes a comparison between them, taking into account their capabilities to simulate near-field wireless communications. In fact, this section is focused on four widely used network DES software: Riverbed Modeler, OMNeT++, ns-2 and ns-3.

2.3.1. *Riverbed Modeler*

Riverbed Modeler[1]—previously known as Opnet Modeler or OPNET—is a commercial network discrete event simulator. It provides a wide range of libraries to model packet networks and it makes use of most of the technologies (e.g. WiFi, WiMAX, ADSL) and protocols (e.g. IP, TCP). Furthermore, it is one of the most popular network simulators and, thus, the third-party support and availability of additional libraries is high.

Riverbed Modeler allows us to exhaustively model the physical characteristics of the wireless communications. Packets are the data chunks processed through the transceiver pipeline of Riverbed Modeler in order to simulate wireless transmissions. The term pipeline is used to outline that Riverbed Modeler processes every wireless packet in a sequence of 14 stages in which the physical characteristics of the wireless link are split. Six of these stages are related to the transmission of the packet: *Receiver Group*, *Transmission Delay*, *Link Closure*, *Channel Match*, *Tx Antenna Gain* and *Propagation Model*. In contrast, the remaining eight stages are related to the reception: *Rx Antenna Gain*, *Received Power*, *Interference Noise*, *Background Noise*, *Signal-to-Noise Ratio*, *Bit Error Rate*, *Error Allocation* and *Error Correction*. The stages are performed sequentially and each stage has available the results of the calculations performed in previous stages. Thus, the *Signal-to-Noise Ratio* stage, for example, has the information about the received signal and interference power estimated in previous stages available (*Received Power*, *Interference Noise* and *Background Noise*) to calculate the Signal-to-Noise Ratio (SNR) and make it available to following stages. In general, the objective of this pipeline workflow consists of calculating the SNR of the packet (*Signal-to-Noise Ratio* stage), then obtaining the Bit Error Rate (BER) of the packet based on the SNR and the modulation used (*Bit Error Rate* stage), estimating the number of wrong bits in the packet (*Error Allocation* stage) and, finally, considering the number of wrong bits to decide if the packet is received correctly or not (*Error Correction* stage).

The simulator comes with several sets of pipeline stages predefined to simulate the most common wireless technologies nowadays such as WiFi or LTE. It also provides a basic set of stages for generic wireless communications such as generic Time Division Multiple Access (TDMA) wireless communication, which could be valid as a basis to model TDMA wireless technologies like GSM.

There are proposals to model RFID in Riverbed Modeler [YAN 09, MAR 13]. The authors of [YAN 09] provide an improved channel model for RFID communications by performing a correct parameterization of the wireless channel

1 http://es.riverbed.com/products/performance-management-control/network-performance-management/network-simulation.html.

and by improving the *BER* pipeline stage with a new BER-SNR curve that takes the fading effect into account. However, no other pipeline stages are modified according to the paper. Instead, the authors of [MAR 13] do not modify any pipeline stage. They study only the link layer protocol of RFID and the wireless coverage is limited with the use of antennas with different diagram patterns.

2.3.2. *OMNeT++*

OMNeT++[2] is a powerful and modular DES software. It consists of modules that can be simple or compound, depending on whether they are atomic or consist of inner modules respectively. The most common way of interaction among modules is by sending and receiving messages via gates and connections. First, gates can be used for sending (output gates) or receiving (input gates) messages. Second, connections can be assigned with transmission properties such as transmission delay or data rate.

There are a huge number of models implemented by individuals and groups made publicly available with open-source licenses. Some of the most important models are provided together in packets known as simulation frameworks. One of the most important is the INET Framework[3] which provides support for the IP family of protocols, wired and wireless link layer protocols, and other popular technologies and protocols. The framework for modeling wireless networks is MiXiM[4]. MiXiM joins and extends several existing simulation frameworks developed for wireless and mobile simulations in OMNeT++. It provides detailed models of the wireless channel (fading, etc.), wireless connectivity, mobility models, models for obstacles and many communication protocols especially at the Medium Access Control (MAC) level. However, MiXiM is now supposed to be deprecated and some of its functionality has been included in INET framework since version 3.0. Finally, it is also worth pointing out the Castalia[5] framework, which is focused on the simulation of Wireless Sensor Networks (WSNs), Body Area Networks (BANs) and networks of low-powered devices. This framework uses the lognormal shadowing model to calculate the average path loss between nodes whose distance is between a couple of meters and hundreds of meters. Moreover, it also provides the alternative of a specific path loss map, e.g. based on real measures, which could be used for BAN and near-field communications.

---

2 https://omnetpp.org/

3 https://inet.omnetpp.org/

4 http://mixim.sourceforge.net

5 http://castalia.research.nicta.com.au/

In [FER 15], the authors introduce a novel simulator to test RFID anti-collision proposals based on OMNeT++ and the Castalia simulation framework. The *Propagation module* of the simulator is responsible for calculating the propagation loss and delay in addition to providing the mechanism to detect RFID collisions.

GreenCastalia[6] [BEN 13] is an extension to the Castalia framework to allow the simulation of protocols and devices that should cope with the energy harvesting typically required in WSNs.

2.3.3. *ns-2*

ns-2[7] is a widely used tool to simulate the behavior of wired and wireless networks. It is an open-source object-oriented DES software organized according to the Open Systems Interconnection (OSI) model. Simulations are based on a combination of C++ and OTcl. In general, C++ is used for implementing protocols and extending the library, and OTcl is used to create and control the simulation environment itself, including the selection of output data. Simulation is run at the packet level, allowing for detailed results.

The MannaSim Framework[8] extends ns-2 to cope with WSNs. In this sense, it introduces new modules for design, development and analysis of different WSN applications. This framework allows for the selection of different types of wireless channels, radio propagation models and antenna models as well as many other physical characteristics of wireless communication.

SensorSim[9] [PAR 00] is a simulation framework developed on top of ns-2 by the US Naval Research Laboratory in order to ease the simulation of sensor networks. It supports different sensor channels which simultaneously support multiple propagation models. Apart from the wireless channel, this framework also focused on other critical aspects of WSNs such as the power model or the energy consumption.

2.3.4. *ns-3*

ns-3[10] is a new simulator, not compatible with ns-2 and built from scratch to replace it. It is entirely built in C++, and OTcl programming language is not used.

6 http://senseslab.di.uniroma1.it/greencastalia

7 http://www.isi.edu/nsnam/ns/

8 http://www.mannasim.dcc.ufmg.br/

9 http://www.nrl.navy.mil/itd/ncs/products/sensorsim

10 https://www.nsnam.org

ns-3 is primarily targeted for research and academic purposes. The large majority of its users focus on wireless/IP simulations that involve models for WiFi, WiMAX or LTE for layers 1 and 2.

Each release of ns-3 is provided with a well documented model library. This model library has support for wireless communication technologies, low-powered wireless communications and up to 15 propagation models that can be extended with other modules.

2.3.5. *Discrete event simulator comparison for near-field communication*

There are multiple surveys that compare these simulators qualitatively [SIN 08, XIA 08, KUM 12], by focusing on the characteristics of each simulator, and quantitatively [KHA 14], by focusing on the performance. From the point of view of modeling near-field wireless communication, none of them provide models to at least simulate the most common near-field communications such as RFID—the near-field version of the RFID specification—and NFC. In fact, there are not even any third-party frameworks that extend the simulators to provide this functionality. Thus, there are two main approaches to integrate the near-field wireless communications in these DES frameworks.

The first approach would be to integrate the near-field wireless communication inside the DES tool by adapting the wireless channel to be able to model near-field wireless communications. Although the physical characteristics of the near-field induction and the far-field propagation are quite different, the features provided to model the far-field propagation could be used to model the near-field induction. In this sense, the four analyzed DES tools present similar capacities. First, Riverbed Modeler has a pipeline of stages to model the wireless channels that can be customized by implementing new stages in C code. OMNeT++ with the Castalia framework and ns-2 with Mannasim or Sensorsim frameworks are focused on Wireless Sensor Networks, but they are also good frameworks to include new channel models. Even ns-3, which is a much more recent simulator, can be extended with new propagation models.

The authors of [ROD 16] present the design of new five pipeline stages of Riverbed Modeler to model the near-field communication between the Eurobalise on the track and the BTM inside the train. Among the modified pipeline stages, the *Closure* stage ensures that the communication is only performed when devices are located very close and the *Power* stage provides an *equivalent* received signal power related to the produced induction that is obtained from an *equivalent* received signal matrix whose values has been decided according to real measures. The design values and validation of the model are carried out with real measures performed in a laboratory with real equipment.

The second approach would be to use *simulation-in-the-loop* or *co-simulation* features of the DES software. Both solutions imply the removal of near-field computation from the DES tool and reliance on measures to be performed in an external device. The simulation-in-the-loop is usually used to connect the simulator to a real device or a real network, whereas the co-simulation is used when the simulator is connected to another simulator. For near-field wireless communications, it would be more appropriate to use the co-simulation option in order to link the DES tool with the electromagnetic simulator. One example of this approach is the COSMO network simulator [ZHA 10] built on top of OMNeT++ and MATLAB to provide an improved indoor wireless simulator. This example could be followed to build a DES simulator with near-field wireless communication capabilities included.

2.4. Conclusions

Nowadays, the increasing availability of low-cost smart objects with wireless near-field interconnection capability represents a growing opportunity to develop smarter and enriched public ITSs. Consequently, a significant research community effort is dedicated to build the necessary tools to evaluate the performance and potential of this massive near-field deployment and their impact on the existent communication infrastructures. System level simulations, such as DES frameworks play a crucial role within this tooling. For complexity reasons, system level frameworks rely on link simulation models accurate enough to capture the essential behavior.

Traditional characterization of near-field communication at the link level relies on two approaches: a theoretical approach normally based on mathematical calculations and another one based on electromagnetic simulators. The last approach has benefited from recent increases in computational power and reached unprecedented levels of simulation capability. With modeling of this low level, the system level simulators benefit from precise communication links and accurate response to the events.

Although most of the commonly used system level simulation frameworks—Riverbed Modeler, OMNeT++, ns-2 and ns-3—target modeling the wireless communication link to measure global end-to-end quality of service performance indicators, their scope is normally far-field technologies. Nevertheless, in recent years, a few initiatives worthy of mention have appeared and they are identified in this chapter.

2.5. Bibliography

[ATZ 10] ATZORI L., IERA A., MORABITO G., "The internet of things: a survey", *Computer Networks*, vol. 54, no. 15, pp. 2787–2805, 2010.

[BEN 13] BENEDETTI D., PETRIOLI C., SPENZA D., "GreenCastalia: an energy-harvesting-enabled framework for the Castalia simulator", *ENSSys'13 Proceedings of the 1st International Workshop on Energy Neutral Sensing Systems*, ACM, New York, pp. 7:1–7:6, 2013.

[CHA 10] CHAKRA S., FARRUKH U., GARCIA B., "Electrical model simulation for a UHF RFID system in near and far fields", *International Journal of Simulation: Systems, Science and Technology*, vol. 11, no. 1, pp. 14–20, 2010.

[DEV 05] DE VITA G., IANNACCONE G., "Design criteria for the RF section of UHF and microwave passive RFID transponders", *IEEE Transactions on Microwave Theory and Techniques*, vol. 53, no. 9, pp. 2978–2990, 2005.

[DOB 12] DOBKIN D.M., *The RF in RFID, UHF RFID in Practice*, 2nd ed., Newnes, 2012.

[EPC 13] EPCGLOBAL, EPC Radio-Frequency Identity Protocols Generation-2 UHF RFID; Specification for RFID Air Interface Protocol for Communications at 860 MHz–960 MHz, EPCglobal Inc., November 2013.

[FER 15] FERRERO R., GANDINO F., MONTRUCCHIO B. *et al.*, "A novel simulator for RFID reader-to-reader anti-collision protocols", *2015 International EURASIP Workshop on RFID Technology (EURFID)*, pp. 59–64, October 2015.

[FIN 04] FINKENZELLER K., *RFID Handbook: Radio-frequency Identification Fundamentals and Applications*, Wiley, 2004.

[GLI 04] GLIDDEN R., BOCKORICK C., COOPER S. *et al.*, "Design of ultra-low-cost UHF RFID tags for supply chain applications", *IEEE Communications Magazine*, vol. 42, no. 8, pp. 140–151, 2004.

[GUB 13] GUBBI J., BUYYA R., MARUSIC S. *et al.*, "Internet of things (IoT): a vision, architectural elements, and future directions", *Future Generation Computer Systems*, vol. 29, no. 7, pp. 1645–1660, 2013.

[HES 09] HESE J.V., SERCU J., PISSOORT D. *et al.*, State of the Art in EM Software for Microwave Engineers, Agilent Technologies Application Note 5990-3225EN, February 2009.

[ISO 13] ISO, IEC, Information Technology – Telecommunications and Information Exchange between Systems – Near Field Communication – Interface and Protocol (NFCIP-1), ISO/IEC, no. 18092, http://cp.literature.agilent.com/litweb/pdf/5990-3225EN.pdf, 2013.

[ISO 16] ISO, IEC, Identification Cards – Contactless Integrated Circuit Cards – Proximity Cards, ISO/IEC, no. 14443, 2016.

[JAN 11] JANKOWSKI-MIHULOWICZ P., KALITA W., "Application of Monte Carlo method for determining the interrogation zone in anticollision radio frequency identification systems", in TURCU C. (ed.), *Current Trends and Challenges in RFID*, InTech, July 2011.

[KAR 03] KARTHAUS U., FISCHER M., "Fully integrated passive UHF RFID transponder IC with 16.7-uW minimum RF input power", *IEEE Journal of Solid-State Circuits*, vol. 38, no. 10, pp. 1602–1608, 2003.

[KHA 14] KHAN M.A., HASBULLAH H., NAZIR B., "Recent open source wireless sensor network supporting simulators: a performance comparison", *2014 International Conference on Computer, Communications, and Control Technology (I4CT)*, pp. 324–328, September 2014.

[KOE 75] KOELLE A., DEPP S., FREYMAN R., "Short-range radio-telemetry for electronic identification, using modulated RF backscatter", *Proceedings of the IEEE; (United States)*, August 1975.

[KUM 12] KUMAR A., KAUSHIK S.K., SHARMA R. *et al.*, "Simulators for wireless networks: a comparative study", *2012 International Conference on Computing Sciences (ICCS)*, pp. 338–342, September 2012.

[LAN 05] LANDT J., "The history of RFID", *IEEE Potentials*, vol. 24, no. 4, pp. 8–11, 2005.

[MAR 13] MARINO F., MASSEI G., PAURA L., "Modeling and performance simulation of EPC Gen2 RFID on OPNET", *2013 IEEE International Workshop on Measurements and Networking Proceedings (M N)*, pp. 83–88, October 2013.

[PAR 00] PARK S., SAVVIDES A., SRIVASTAVA M.B., "SensorSim: a simulation framework for sensor networks", *Proceedings of the 3rd ACM International Workshop on Modeling, Analysis and Simulation of Wireless and Mobile Systems*, MSWIM'00, ACM, New York, pp. 104–111, 2000.

[ROD 16] RODRIGUEZ L., PINEDO C., LOPEZ I. *et al.*, "Eurobalise-Train communication modelling to assess interferences in railway control signalling systems", *Network Protocols and Algorithms*, vol. 8, no. 1, pp. 58–72, 2016.

[SIN 08] SINGH C.P., VYAS O.P., TIWARI M.K., "A survey of simulation in sensor networks", *2008 International Conference on Computational Intelligence for Modelling Control Automation*, pp. 867–872, December 2008.

[SWA 03] SWANSON D.G., HOEFER W.J., *Microwave Circuit Modeling Using Electromagnetic Field Simulation*, Artech House, London, 2003.

[UNI 08] UNISIG, FFFIS For Euroloop v2.3.0, ERTMS/ETCS Class 1, SUBSET, no. 044, February 2008, available at: http://www.era.europa.eu/Document-Register/Pages/Set-2-and-3-FFFIS-for-Euroloop.aspx.

[UNI 12] UNISIG, FFFIS For Eurobalise v3.0.0, ERTMS/ETCS Class 1, SUBSET, no. 036, February 2012, available at: http://www.era.europa.eu/Document-Register/Pages/Set-2-and-3-FFFIS-for-Euroloop.aspx.

[WAN 06] WANT R., "RFID explained: a primer on radio frequency identification technologies", *Synthesis Lectures on Mobile and Pervasive Computing*, vol. 1, pp. 1–94, January 2006.

[XIA 08] XIAN X., SHI W., HUANG H., "Comparison of OMNET++ and other simulator for WSN simulation", *2008 3rd IEEE Conference on Industrial Electronics and Applications*, pp. 1439–1443, June 2008.

[YAN 09] YANG D., LIU W., "The wireless channel modeling for RFID system with OPNET", *2009 5th International Conference on Wireless Communications, Networking and Mobile Computing*, pp. 1–3, September 2009.

[ZAN 14] ZANELLA A., BUI N., CASTELLANI A. *et al.*, "Internet of things for smart cities", *IEEE Internet of Things Journal*, vol. 1, no. 1, pp. 22–32, 2014.

[ZHA 10] ZHANG Z., LU Z., CHEN Q. *et al.*, "COSMO: co-simulation with MATLAB and OMNeT++ for indoor wireless networks", *2010 IEEE Global Telecommunications Conference (GLOBECOM 2010)*, pp. 1–6, December 2010.

3

Trace Extraction for Mobility in Civil Aeronautical Communication Networks Simulation

The mobility pattern of nodes in a wireless network has a strong impact on the communication technologies that can be deployed. It is therefore logical that accurately simulating this mobility is the first step toward simulating the whole communication system. In the case of a civil aircraft, the mobility in a given airspace is constrained by air authority regulations in addition to aircraft capacities. Furthermore, general air traffic is shaped both by regulations and economic considerations. These, plus the physical characteristics of aircraft (e.g. speed, altitude), create a specific mobility pattern for each airspace considered.

In network simulation, the mobility of nodes can be seen in one of two ways. It can first be considered as an input to the network model, with stochastic models or pre-recorded traces. It can also be seen as a process that both impacts the communications and is impacted by them. In this last case, the mobility must be implemented by simulating the behavior of autonomous network aware agents that must communicate in order to perform a common goal (e.g. reaching their destinations while respecting safety measures). As far as aircraft communication simulations are concerned, researchers tend to use the former point of view owing to the fact that current regulations on air traffic do not offer much choice of trajectory to the individual aircraft.

In the domain of aeronautical communications, network simulation is often used to test and design new solutions. These solutions range from satellite communications for oceanic zones as in [PIR 13] to ad hoc networks with aircraft as nodes, called Aeronautical Ad Hoc Networks (AANET).

Chapter written by Fabien GARCIA and Mickaël ROYER.

In this chapter, we will first discuss the current traffic regulations in different airspaces in order to lay out the constraints that exist on aircraft movement. In section 3.2, we will discuss different types of mobility models and their respective merits. In section 3.3, we will discuss traffic trace extraction, enhancement and filtering. Finally, we will talk about new developments on cooperative trajectories' studies that seem to be a new trend.

3.1. Traffic regulations

Commercial civil aircraft do not fly in a "straight path"[1] between their departure and arrival airport. They instead use routes defined by air traffic regulation authorities in order to share the airspace between civil and military organizations and to avoid aircraft coming too close to one another or even colliding.

Airspace is divided into several regions. The first distinction that can be made is between military and civil airspaces; in this chapter, we will only look into commercial civil aircraft traffic and hence will develop the rules that apply to civil airspace. We will particularly look at two different airspaces that present different characteristics due to the available means of traffic control: continental airspace and oceanic traffic. For the latter, we will look into the North Atlantic Airspace which is one of the busiest the region in the world as far as air traffic goes.

3.1.1. *General airspace*

The civil airspace above land is divided into three dimensional blocks of different *classes*. Each class of airspace imposes different rules on the air traffic that can go through it. These classes, as defined by the International Civil Aviation Organization (ICAO), range from class A to G. Not all classes are implemented in each country, but their general meaning is the same all over the world.

Classes A through E constitute controlled airspace within which Air Traffic Control (ATC) services are provided. Classes F and G are uncontrolled airspace where ATC services are not maintained. The different classes allow different types of flights to go through, from Visual Flight Rules (VFR) traffic consisting of aircraft piloted by looking out of the cockpit to Instrument Flight Rules (IFR) which consist of aircraft piloted only through on-board instruments.

Commercial aircraft are all piloted applying IFR and in controlled airspace. They follow routes specified by their *flight plan* as a succession of predefined pathways that generally intersect at radio navigation stations. An example of a resulting route is given

1 The concept of straight path on the surface of a sphere is here used for "as the crow flies"; see great circle interpolation in section 3.3.1.

in Figure 3.1 Aircraft following the same route must be separated by a certain distance horizontally or vertically. This separation can be seen as a three dimensional version of safety distances on highways. These characteristics of commercial aeronautical traffic lead to similarities with car traffic on highways, and emphasize the common points between Vehicular Ad Hoc Network (VANETs) and AANETs.

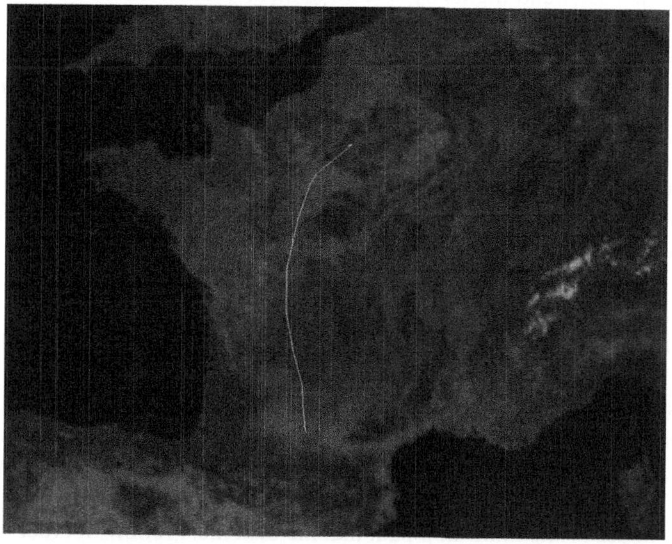

Figure 3.1. *Example of a flight plan between Toulouse and Paris in France. For a color version of this figure, see www.iste.co.uk/hilt/transportation.zip*

3.1.2. *North Atlantic airspace*

The North Atlantic Airspace is a zone extending roughly from 10°N to the North Pole and from 60°W to 10°W. It contains five areas named after the ATC center responsible for enforcing regulations in them: Reykjavik, Shanwick, Gander, New York and Santa Maria. As [EUR 16] states: "The airspace of the North Atlantic which links Europe and North America is the busiest oceanic airspace in the world. In 2012 approximately 460,000 flights crossed the North Atlantic." Due to this aircraft density and the lack of radar equipment in this zone, ATC must enforce strict separation rules between aircraft by assigning routes to them and having each aircraft report its position periodically (one report every hour or each degree of longitude covered).

Most of the traffic in the North Atlantic Airspace is organized into two major flows: a westbound flow departing Europe in the morning and an eastbound flow departing North America in the evening (UTC time). The first flow crosses the 30°W longitude between 11:30 and 19:00 (UTC) and the second flow crosses the same longitude between 01:00 and 08:00 (UTC). This Organized Track System (OTS) is the result of trying to satisfy passenger demands and airport noise reduction considering the time zone differences. Furthermore, it allows easy separation of westbound and eastbound flights by spreading them in different time slots. The set of westbound and eastbound tracks available for flying is published daily by ATC centers (Shanwick and Gander, respectively). It is chosen so as to allow a maximum number of aircraft to follow them while guaranteeing the most economic flight conditions (considering winds and aircraft preferred altitude). Finally, it is to be noted that these tracks are not mandatory, and the International Civil Aviation Organization (ICAO) considers that half of aircraft follow them as of 2016 [EUR 16], aircraft not following the tracks must nevertheless expect less than optimal flight altitude and frequent rerouting to ensure aircraft separation rules within the NAT.

3.2. Mobility for network simulation

In classical, wire-based infrastructure networks, the relative position of the different pieces of equipment only impacts the physical delay between them and as such is often only modeled through this parameter. In wireless networks, the relative position of the nodes also impacts, for instance, the interference between the signals of different nodes or the multi-hop path that the data must follow in order to reach its destination. Positioning the nodes accurately therefore becomes of prime importance if we want a simulation to be realistic.

Several types of mobility models have been proposed and used in the literature, which will be briefly classified in section 3.2.1. Then, we will compare them in section 3.2.2 in terms of their merits in accurately modeling the behavior of mobile nodes in different aeronautical wireless networks.

3.2.1. *Types of mobility models for AANETs*

Surveys of mobility models in mobile ad hoc networks existing in the literature, as for example [ROY 11, CAM 02, BAI 04], consider mobility models that define the speed and position of mobile nodes in a random way. The classification of such models is then carried out according to the rules that govern the successive updates and those that relate updates to the position of different nodes.

A generally accepted classification distinguishes individual mobility models that determine the mobility of nodes independently from other nodes, and groups mobility models that define correlations in the updates of a node position and speed as well as those of other nodes.

Such a classification generally encompasses only *random mobility models*, and these models define the mobility through a stochastic approach. They are generally easy to compute and hence do not impair the performance of network simulation. Furthermore, as mentioned in [BAI 04], the fact that MANETs are not yet a common occurrence in everyday life makes it hard to gather traces of real node mobility and impose the use of random mobility models. In aeronautical communications, traces of real mobility do exist[2]. Wireless network simulations in the aeronautical domain can therefore be carried out with not just realistic, but real mobility patterns for the nodes.

In the field of network simulations (and particularly in the domain of MANETs), random models are thus more often used than traces. The most common individual models include *random waypoint*, *random direction* and *random Gauss-Markov*. In the random waypoint model, nodes choose a random waypoint, a speed and a pause time (according to stochastic laws given as parameters of the model). It then goes toward the waypoint at the constant speed chosen and pauses there for the given time. In aeronautical network simulations, the pause time is of course fixed to zero. The random direction model involves choosing a random direction instead of a next waypoint and going in this direction until the nodes reach the limit of the simulation. In the Gauss-Markov model, the motion of the nodes depends on their current speed and direction; at the beginning a random direction and speed are chosen (again according to random distributions given as parameters), and at each update the new speed and direction are computed by applying a deviation on the previous values. This last model gives much smoother movements of the nodes than the two previous models which generate sharp turns, it is therefore viewed as more realistic. An example of two nodes moving according to this model is given in Figure 3.2, which can be compared to the other examples of real aircraft trajectories throughout this chapter, to asses the unrealistic character of such movement applied to commercial aircraft.

3.2.2. *Comparison of mobility model types*

In this section, we will not discuss the merits of each individual mobility model but rather try and emphasize the advantages and drawbacks of using real traces compared to random mobility models. Indeed, the availability of mobility traces might lead one to think that random models are no longer useful but care must be taken in this. First of all, using purely random mobility models can lead to faster simulations which can be an enticing feature in the early stage of development of a new communication system. Second, the number of different traces available is an important point. Network simulation uses a statistical approach to testing a system, it is therefore important to have statistically representative mobility patterns as an input

2 It must be noted that such traces also exist for VANETs, where nodes are ground vehicles, but we will concentrate on aircraft in this chapter.

to the simulation. In the case of mobility traces, this comes at the expense of a sufficiently large number of traces fed to the simulation.

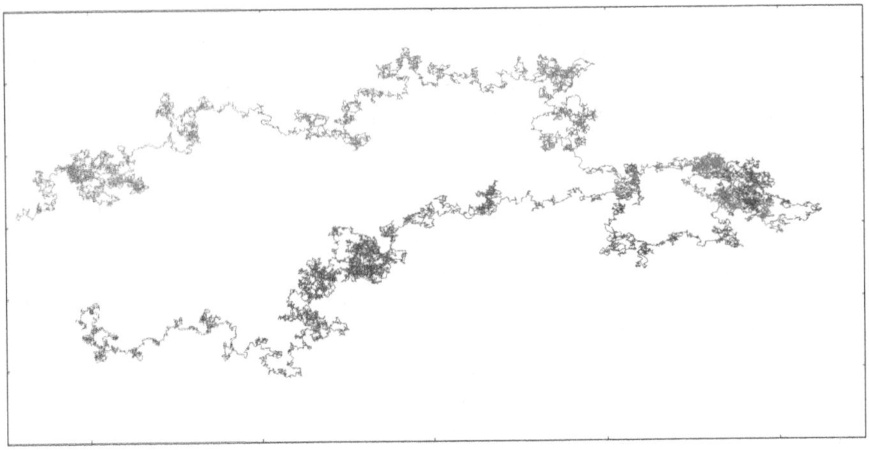

Figure 3.2. *Example of two nodes moving according to the Gauss-Markov model. For a color version of this figure, see www.iste.co.uk/hilt/transportation.zip*

The choice between traces and random mobility models therefore appears as a classical trade-off between the time and resources available for simulating the system (including the time to gather and pre-process the traces and the available data storage resources) and the accuracy of the simulation.

It should be noted though that another option might lead to the advantages of both solutions (i.e. realistic mobility and variability in the input mobility patterns) with the advent of a mobility model generator that would include current air traffic regulations and realistic flight plan distribution in order to reconstruct realistic traffic. This kind of solution is already used in a terrestrial vehicle mobility simulation (see, e.g., [SOM 11]) and could allow for further developments in mobility models such as those presented in section 3.4.

3.3. Example of mobility trace extraction

In this section, we will present an example of trace extraction for aeronautical communication simulation over the north Atlantic airspace. As stated earlier, this airspace is interesting both for the heavy traffic that passes through it and for the lack of infrastructure that prevents any cell-based communication mechanisms. This type of trace extraction has been previously used in several studies like [PIR 13, BES 11] or [VEY 16]. The data used for these studies was taken from the Data Demand Repository of Eurocontrol available from the OneSky Online website[3]. Eurocontrol is the "European Organisation for the Safety of Air Navigation" and acts as a central organization for coordination and planning of air traffic control for all of Europe. Their Demand Data Repository provides pan-European air traffic forecasts and historical data, and hence contains all flights going to, coming from or passing over Europe, making it a prime source of traces of aircraft movements.

Extraction of mobility traces for network simulation can be decomposed into three steps, each of which is covered in the following sections. These three parts consist of extracting the information, filtering it so as to target the simulation and then improving the traces so as to use them in network simulation.

3.3.1. *Extraction of information*

Extracting the information from traces can be as simple as parsing the data and translating it to a format compatible with the network simulator that will be used. Care should be taken to gather traces that are sufficiently detailed. An example of insufficiently detailed sources would imply using the departure and arrival airport only and interpolating between them using the shortest route. This can lead to large errors in the trajectories. An example of such an error is shown in Figure 3.3 where the real trajectory is to the north. These differences can also be seen in continental airspace and are due to the routes that have been presented in section 3.1. As mentioned previously, in section 3.1.2, in the case of NAT airspace, the routes change daily according to weather conditions as a longer distance route might lead to shorter flights depending on the wind.

3.3.2. *Traces filtering*

The most loaded day recorded by Eurocontrol in 2015 was the 28th of August and had 34734 flights. The least loaded day was the 25 of December and had 12508 flights. For this last day, the maximum number of aircraft in flight at the same time[4] was 739. Even for such a low traffic day, the number of aircraft in flight at a given time can lead to slowdowns in simulation. As we are generally interested in only a portion of the traffic, filtering is an important part of trace extraction and use.

The most obvious type of filters are geographic ones which will select the traffic in a given zone or according to the altitude in order to select aircraft in a given state of flight (e.g. approaching or leaving an airport or en route). Filters that select only aircraft that cross a certain meridian or latitude are good as they quickly select

3 See: https://ext.eurocontrol.int/

4 It is also called the Peak Instantaneous Aircraft Count (PIAC).

aircraft that travel in a general direction over a certain area and leave fewer aircraft to be processed for finer filters. These are not the only type of filters though, filtering on the *callsign*[5] might allow us to only include commercial aircraft from a given company (e.g. to test connectivity in ad hoc networks composed of aircraft of one company only). Another type of filter includes aircraft type (e.g. in the case where the feature tested could not be transported in some smaller aircraft, such as bulky satellite antennas) or departure and arrival airport.

Figure 3.3. *Trajectory error with interpolation from departure airport to arrival airport. For a color version of this figure, see www.iste.co.uk/hilt/transportation.zip*

Selecting the right filters can drastically decrease the number of flights to be processed during simulation and help target only the relevant aircraft. For example, on the 25th December 2015 if we apply a filter for aircraft crossing the $30°W$ meridian (which selects most aircraft in the NAT airspace), the number of aircraft to be treated goes down to 1025. After applying a filter to select all aircraft above an altitude of 5000 ft and in the $90°W$ to $10°E$ and $23.5°N$ to $70°N$ zone, the number of aircraft left is 1017 with a maximum of 338 aircraft in flight at the same time.

3.3.3. *Enhancing traces*

Trajectories gathered from such sources as those discussed above provide the positions of aircraft at irregular times from 1 minute to 1 hour apart. These vary

5 The callsign is the unique ID given by civil aviation authorities to an aircraft or flight.

according to the technologies used for reporting the positions. Over land, radar is often used and reports the position roughly each minute. Over the north transatlantic corridor, the reports are done by the aircraft each hour or each degree of longitude traveled (whichever happens first).

As network simulation generally requires time steps in the order of the millisecond, we should interpolate the original data. This interpolation should be done following the rules for navigation that lead to shortest routes on a sphere called *orthodromy*. This kind of route does not follow a constant heading (i.e. the angle formed by the route and meridians) and must be computed iteratively from the current position and the intended destination. For the purpose of precision, it is a good idea to interpolate along orthodromies when the position reported is far away and to use loxodromy when they are sufficiently close, as this is generally the way navigation is done.

It should be noted that the two types of interpolation following the shortest route or a constant heading route are entirely different from linear interpolation in the latitude/longitude two-dimensional plane. The aircraft can be thought of navigating on the surface of a constant altitude sphere during most of their flight (at least in NAT airspace) and, as the portion of the sphere covered is quite large, it cannot be accurately projected on to a plane.

An example of what the traffic can look like is given in Figure 3.4. This figure shows 1017 flights crossing the $30°W$ meridian at cruise altitude limited to the geographic region $90°W$ to $10°E$ and $23.5°N$ to $70°N$.

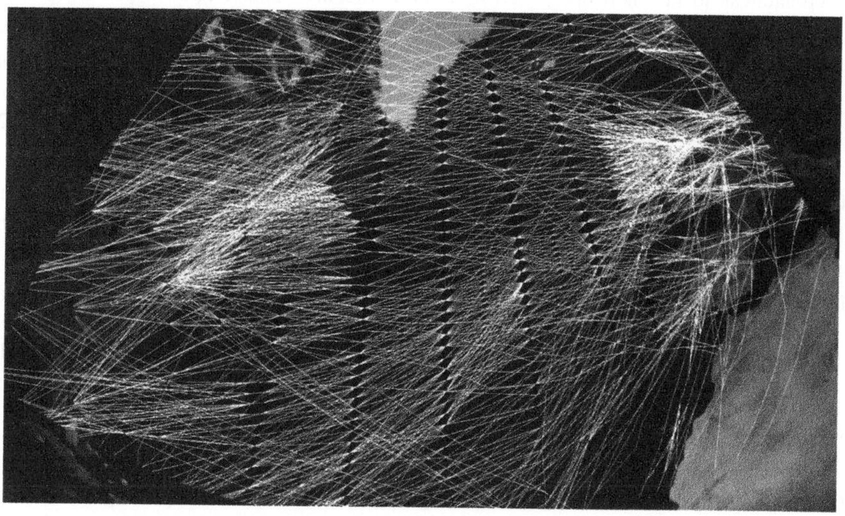

Figure 3.4. *North Atlantic Track airspace traffic on 2015/12/25. For a color version of this figure, see www.iste.co.uk/hilt/transportation.zip*

3.4. Toward cooperative trajectories

The mobility models presented up until now in this chapter are seen as an input to the network simulation. However depending on the applications envisioned, the communication itself could lead to a change in mobility. One typical example would be cooperative aircraft trajectory planning, where aircraft exchange position information with those around to cooperatively ensure that no aircraft will go too near another aircraft. In such a system, communications will have an influence over mobility, and as in other examples across this chapter, the mobility (through varying distances between aircraft) will influence the communications.

This bidirectional dependency was successfully taken into account in the VEINS framework [SOM 11] for terrestrial communications, and will be crucial toward the development of cooperative communicating autonomous agents aboard aircraft. The development of such systems could lead to a breakthrough in Air Traffic Control systems and help increase safety and efficiency in zones where radar coverage is not possible (e.g. NAT airspace).

3.5. Bibliography

[BAI 04] BAI F., HELMY A., "A survey of mobility models", *Wireless Ad-Hoc and Sensor Networks*, University of Southern California, 2004.

[BES 11] BESSE F., PIROVANO A., GARCIA F. *et al.*, "Interference estimation in an aeronautical ad hoc network", *DASC 2011 IEEE/AIAA 30th Digital Avionics Systems Conference*, Seattle, pp. 4C6-1–4C6-11, October 2011.

[CAM 02] CAMP T., BOLENG J., DAVIES V., "A survey of mobility models for ad hoc network research", *Wireless Communications & Mobile Computing (WCMC): Special Issue on Mobile Ad Hoc Networking: Research, Trends and Applications*, vol. 2, pp. 483–502, 2002.

[EUR 16] EUROPEAN AND NORTH ATLANTIC OFFICE OF ICAO, "North Atlantic Operations and Airspace Manual", International Civil Aviation Organization (ICAO) document from the EUR/NAT office, ICAO reference: NAT Doc 007 V.2016-1, p. 207, 2016.

[PIR 13] PIROVANO A., GARCIA F., RADZIK J., "Capacity dimensioning for aeronautical communications in North Atlantic Corridor", *KACONF 2013, 19th Ka and Broadband Communications, Navigation and Earth Observation Conference*, Florence, p. 8, available at: http://www.kaconf.org/CD2013/papers/Ka_3/63.pdf, October 2013.

[ROY 11] ROY R.R., *Handbook of Mobile Ad Hoc Networks for Mobility Models*, Springer, Boston, 2011.

[SOM 11] SOMMER C., GERMAN R., DRESSLER F., "Bidirectionally coupled network and road traffic simulation for improved IVC analysis", *IEEE Transactions on Mobile Computing*, vol. 10, no. 1, pp. 3–15, January 2011.

[VEY 16] VEY Q., PIROVANO A., RADZIK J., "Routing protocol assessment for AANETs", *AEGATS '16, Advanced Aircraft Efficiency in a Global Air Transport System*, Paris, http://oatao.univ-toulouse.fr/16026/ mentions, pp. 1–9, April 2016.

4

Air-Ground Data Link Communications in Air Transport

The current evolution of the civil aviation industry shows a drastic increase in data exchanges between on-board and ground systems. These data are related to safety, eco-friendliness and economic purposes. The overall set of solutions, including the communication system and the applications, is known as the aeronautical data link. Regarding the considered airspace, different communication systems can be used. Some of these recent systems, such as VDL (VHF Data Link), are based on the line-of-sight links between aircraft and ground stations, thus limiting their deployment to the continental domain. In oceanic areas, satellite-based systems are proposed as the main solution for future aeronautical data link communications known as AMSS (aeronautical mobile satellite service). Both systems are intended to support very different types of services with mobile nodes. In this context, traffic characterization, communication architecture and protocols have to be explored and validated. These are the fields in which a simulator becomes handy, allowing the validation of techniques and algorithms.

4.1. Introduction

4.1.1. *Context*

Aeronautical communication embraces a wide spectrum of usage, from the passengers' desire for on-board Wi-Fi with Internet connectivity and airline data collection for the cost efficiency of aircraft operation, to safety-related communication between the pilot and the controller for managing the air traffic. All these applications require wireless communication means between the aircraft and

---

Chapter written by Christophe GUERBER, Alain PIROVANO and José RADZIK.

the ground, with different Quality-of-Service requirements. Moreover, the environment and conditions in which these communications take place vary widely during the flight: from an on-ground low-speed and dense area on the airport runway to a high-altitude environment at high speed, known as the en-route airspace. The simulations described in the following sections consider those technologies that are dedicated to en-route airspace for safety of life communications, mainly supporting the communications taking place between a controller and a pilot [BEN 13].

Currently, these communications are mainly based on voice analog radios, providing a quite intuitive yet error-prone communication path between controllers and pilots: misunderstandings, inefficiencies and errors in executing clearances, with no automation possible. The allocated spectrum for these communications includes both VHF band from 118 to 137 MHz for continental areas and HF band in remote and oceanic airspace.

The steady increase in aircraft traffic demand all over the world pushed the air traffic industry to look for safety and efficiency improvements. They will be achieved, *inter alia*, through the use of digital communication technologies, which enable the increased quality and efficiency of the communication path and a higher level of automation.

For several years now, the International Civil Aviation Organization has been working on the development and deployment of the Aeronautical Telecommunication Network (ATN). This worldwide internetwork will gather all the air traffic management stakeholders together, facilitating the sharing of operational information and supporting of near real-time applications for the control of air traffic. This is to say that aircraft are part of this internetwork and should be provided with digital air-ground telecommunication means in the different airspace.

For this purpose, VHF band is a good candidate for supporting these communications with achievable high availability, low delays and high throughput in continental areas. After a technology selection process, it is the VHF Data link mode 2 technology that has been elected as the air-ground sub-network for continental areas and is currently deployed in several regions of the world. VDL mode 2 is the subject of the first simulation model presented in the following sections.

For oceanic and remote airspace, satellite-based communication seems the only technology to provide the Quality-of-Service suitable for the above-described applications and is the subject of the second simulation model to be described hereafter.

4.1.2. *OMNeT++*

OMNeT++ [OMN 16] is a discrete event simulation system based on C++, which mainly focuses on communication networks and distributed systems. This is an open-source and research-oriented framework. It enables large-scale simulation with hierarchical models. A discrete-event simulation is a chronological sequence of the occurrences of events. This approach requires an event list to be maintained, insertion into and deletion from it to be enabled, the simulation clock to be handled and utilities to generate random numbers from common probability distributions to be provided. Varga [VAR 08] gives detailed information on OMNeT++ and compares it with other frameworks dedicated to network simulations.

4.2. Continental air-ground data link communications and VDL mode 2

4.2.1. *Communication system*

VDL mode 2 uses the same protocols as those of X.25 at the interface between the aircraft and the ground station, although no requirements apply to the supporting ground wide-area network. VDL mode 2 should provide a reliable connection-oriented network service between an on-board ATN router and a ground ATN router. An additional connectionless service is also provided at the link layer level but has currently no standardized use. Air-ground connectivity is provided through several ground stations, building a mobile network that manages mobility through handing the aircraft over the different stations across its radio coverage. The same radio channel is operated by different ground stations, easing the transition from one station to the other. Ground stations may operate several channels, in which case tuning parameters are provided by the ground stations. The mobile airborne stations will hand over from one station to another by applying the so-called "make before break" paradigm (soft handoff). Managing these handoffs in an efficient way is of prime importance to maintain the connectivity between air and ground routers. An aircraft is required to be able to manage handoffs on its own; however, options are provided for the ground to perform handoffs or to require aircraft to do so, for example, to manage channel load. The latter is also essential to maintain an acceptable Quality-of-Service, especially for maintaining low transit delays.

From a physical layer viewpoint, VDL mode 2 is operated in the aeronautical VHF band and more precisely in the upper channels of this band, between 136.900 and 136.975 MHz (see ICAO Annex 10 vol. 4 [ICA 07]). VDL mode 2 uses the same channel spacing as that of the voice channel, namely 25 kHz channels, and a differential eight-phase shift keying modulation that provides 31,500 bps per channel. Channel coding consists of a Reed–Solomon block coding, with each block being interleaved to spread error burst. Transmission thus consists of a frame with a start of transmission identifier and the length of the transmission to allow the de-interleaving stage to perform its job. This frame acts as a container for the link layer protocol frames that may be grouped into a single access to the channel. The maximum length of a transmission is limited by the transmission length encoded on 17 bits (131,071 bits). A bit scrambling stage with a fixed initialization value is performed before the data are provided to the modulation stage. Additional modulation/demodulation optimization techniques are used to enhance bit error rate. The requirement here is a BER of 10^{-4} at the output of the physical layer inside the intended coverage.

Channel access follows the CSMA p-persistent rule: for a transmission to take place, the channel is first sensed for idle/busy state. If busy, the station will persist in listening to the channel to wait for it to become idle. Each time the channel is tested idle or has just become idle, the station transmits with probability p and waits for a slot time with probability $1-p$. After each time slot, the process is started again, until the transmission is performed, or the maximum number of transmission attempts is reached, or the maximum channel access timer expires.

The link layer protocol is a derived version of ISO protocol HDLC, renamed Aviation VHF Link Control. Key characteristics include the use of a modified selective reject frame for selectively rejecting and acknowledging frames intended to minimize the number of unnecessary retransmissions. Acknowledgments are delayed to allow grouped acknowledgments and piggybacking of the acknowledgment into an information frame. Information frames are sent with a transmission window of 4 to accelerate the transmission in case of bursts of data (e.g. several segments of data). No priority, neither within transmission nor between transmitting stations, is defined. On top of the AVLC protocol, an 8208 network protocol allows large data that would not fit into the maximum frame length to be segmented and several virtual circuits inside a single link layer connection to be multiplexed. Flow control may also be achieved here by delaying acknowledgments.

From the viewpoint of link layer connection management, ground stations announce themselves through the sending of an identification frame containing a protocol parameter to be used as well as the DTE address of the reachable routers. Mobile stations are expected to listen for these identification frames to discover the available ground stations and establish the initial link. Handoff and channel load management requires both air and ground to gather information on the peer station. Signal quality measurements on each received transmission and a few timers allow the stations to acquire a reasonably good knowledge about the surrounding other stations to try to manage handoffs in an efficient way. On the ground side, stations sharing the same knowledge are said to belong to the same VDL mode 2 ground system. When the conditions require the aircraft to perform a handoff or if the current ground station requires so, the aircraft will establish a link layer connection with a new station before disconnecting the old link and reestablish all the necessary virtual circuits. Handoffs will happen in regions where radio coverage of at least two ground stations from the same ground system overlaps. There is no requirement for these ground stations to be synchronized for channel access. Handoffs between two separate ground systems are treated as the first link and require an explicit disconnection of the old link. Optionally, the handoff may require the mobile station to retune its radio on another channel. Different deployment scenarios exist in a multi-frequency operation.

4.2.2. *Dimensioning parameters and bottlenecks*

As explained in the previous part, the VDL mode 2 architecture covers the functionalities provided from the physical layer to the first subpart of the network layer. Of course, the number of dimensioning protocols parameters and bottlenecks is potentially high, considering the relevant layers, particularly in a wireless communication environment. Considering outgoing packets from an end system, a first bottleneck is met in the DLS (Data Link Service) sublayer with the AVLC protocol. In order to ensure point-to-point reliability with flow control, AVLC uses a sliding window with a default size of four frames. Hence, as shown in Figure 4.1, the packets have to be potentially enqueued until previously sent frames are acknowledged. This point is particularly relevant in the ground station, where several DLE (Data Link Entity) may be present, that is, one for each connected aircraft. In the VME (VDL Management Entity) sublayer, each LME manages the AVLC connection between an aircraft and the ground station. Hence, for a given traffic load generated by the upper layers, AVLC parameters have to be tuned in order to ensure an efficient flow control while avoiding congestion in queues. The important parameters are:

– the window size k;

– the delay before ACK T2;

– the maximum number of bits in any frames N1 (default: 8,312 bits);

– the maximum number of transmissions N2 (default: 6).

The delay before retransmission T1 is computed and updated during the different connections as a function of several parameters. Notably, the TD_{99}, that is, the observed transaction delay (from application layer to application layer) for 99% of packets is one of these parameters.

The MAC sublayer is also driven by a set of parameters that have to be efficiently tuned. As explained in the previous part, this sublayer is based on the CSMA p-persistent protocol in order to prevent collisions between frames sent by the different nodes.

The main relevant parameters are:

– the probability p to transmit if the channel is idle (default: 13/256);

– the interaccess delay timer *TM1* between two attempts (default: 4.5 ms);

– the maximum number of access attempts *M1* (default: 135).

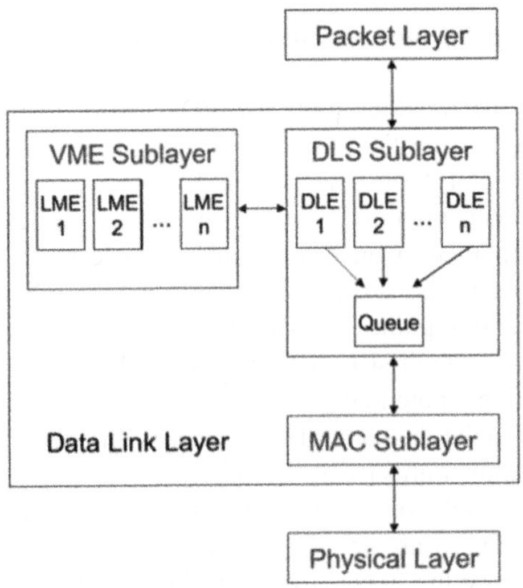

Figure 4.1. *VDL mode 2 layers and entities*

Considering the asymmetric topology given by a group of aircraft covered by a single ground station, it has to be verified that the protocols and different mechanisms operate in a fair way. For instance, the mean waiting time in the queue that feed the MAC sublayer has to be approximately identical in the aircraft and the ground station.

Finally, the physical layer also includes important parameters. In our context, mainly dedicated to CSMA and AVLC protocols, we consider the maximum length of the physical frame (131,071 bits) that allows several MAC frames to be aggregated. Furthermore, it has to be underlined that the channel capacity (31.5 kbit/s) at the physical layer also represents a potential bottleneck.

4.2.3. *Simulation model*

The main goals of the simulation model are to assess the performances of VDL mode 2, considering the CSMA p-persistent and AVLC parameters, and to provide pedagogic tools to students. As shown in Figure 4.2, the studied model represents a geographical zone covered by a single VGS (VHF Ground Station). The number of visible aircraft is one of the model parameters.

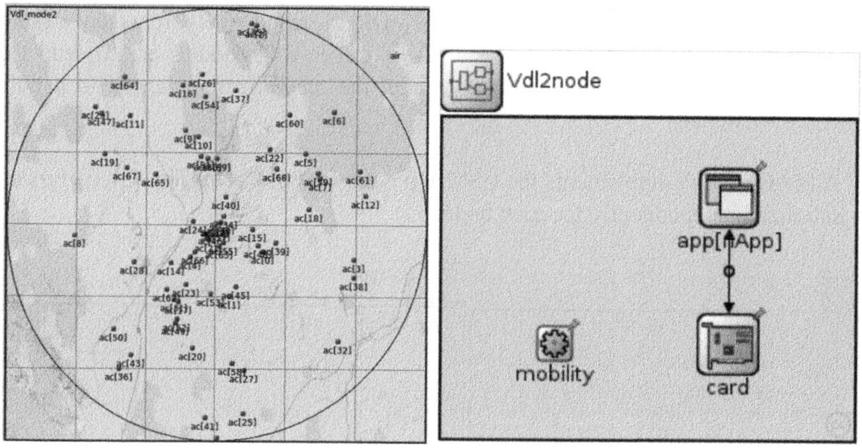

Figure 4.2. *VDL mode 2 topology and node models*

The model includes two types of node: one VGS (VDL Ground Station) and several aircraft. Another module named medium through which all the messages are sent is also included. Both the ground station (Ground Station module) and the aircraft (Aircraft module) extend the Vdl2node module presented in Figure 4.2. As the behavior of a node is complex, the functionalities of the Vdl2node are split into several submodules.

These submodules are:

– Application (app): this module allows the generation of messages from aircraft or from GS application layers. The generated messages have a random length using a uniform distribution between a minimum length of 32 bytes and a maximum length of 265 bytes. And for a single DLE, the time between generated messages has a random value, using an exponential distribution, with a mean value of 40 s (by default);

– VDL card: the VDL card module implements all the VDL mechanisms, from the physical layer to the data link layer. It is a complex component that contains several submodules;

– Mobility module: this represents the position of a node and provides several useful functions to calculate distances between nodes. Considering the current objectives of the simulation model, node positions are static. They are precomputed and read from a file during the initialization of simulations.

As the VDL card is the core of the model, the following paragraphs offer an insight into its submodules shown in Figure 4.3.

The physical layer modules VDL_rx and VDL_tx, respectively, represent a radio receiver and a radio transmitter. The aims of these two modules are to send (to receive) frames to (from) other nodes through the medium module and to handle collisions of signals.

A MAC module implements the CSMA-p persistent protocol. The value of p can be initialized independently for each node.

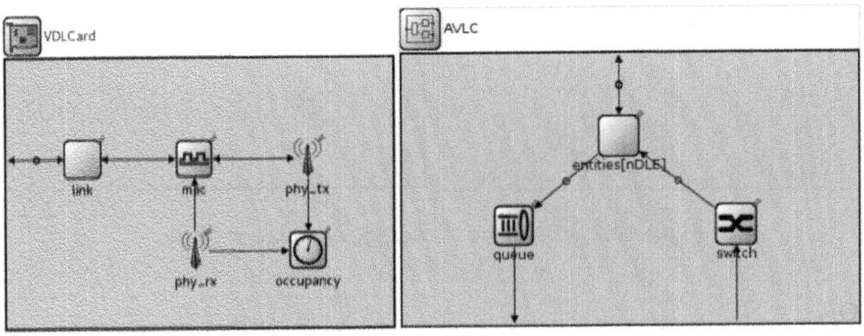

Figure 4.3. *VDL card model and AVLC module*

The AVLC module, presented in Figure 4.3, implements the AVLC communication protocol. It is a compound module. The nDLE submodule represents the AVLC DLE. The Queue module models the queue that stands between the DLEs and the MAC sublayer. This queue plays a very important role in the model. As the CSMA-p protocol senses the channel before sending a packet from the queue, the time spent in the queue may not be negligible. Therefore, a correct management of the queue by DLEs is crucial in order to avoid sending outdated frames and congesting the channel. And, as the radio frames are broadcast to all the nodes, the Switch submodule filters the received frames using the destination address indicated in the frames header. It has to be noted that, as the connection phase is not modeled, there is no module representing the LMEs (Link Management Entities).

The Medium module is designed to broadcast messages sent by nodes and to eventually apply bit corruption and packet losses on the VHF channel. Furthermore, this module sets a propagation delay for each node according to the distance between senders and receivers.

4.2.4. *Analysis of simulation results*

We illustrate here the different types of potential results with some examples.

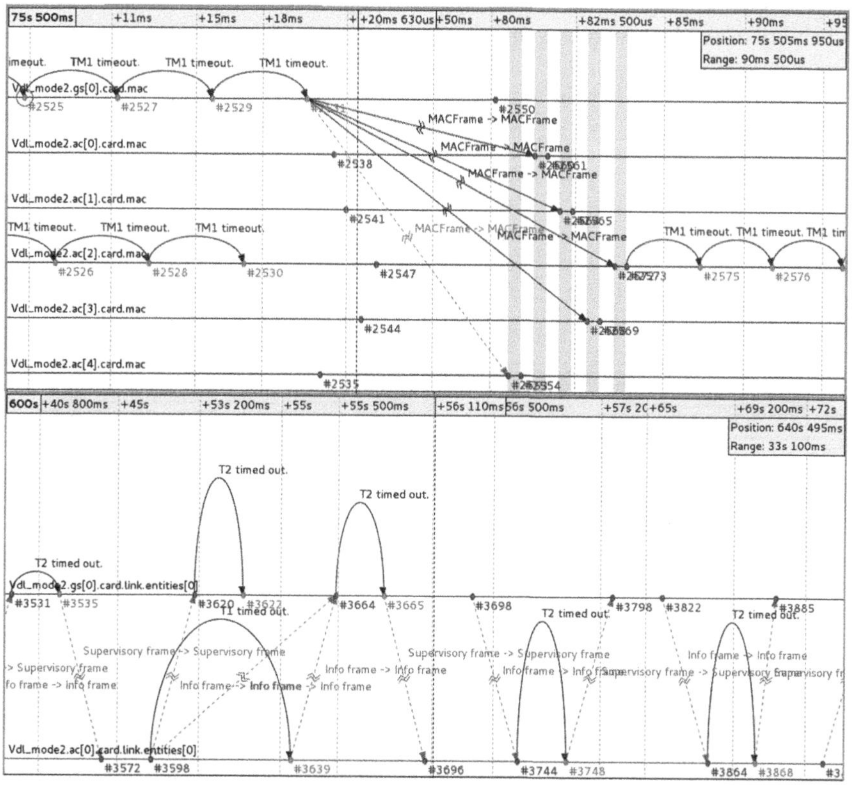

Figure 4.4. *Sequence charts of MAC and AVLC layers. For a color version of this figure, see www.iste.co.uk/hilt/transportation.zip*

The simulation results help us to assess the performances of the VDL mode 2 communication system, considering the number of aircraft covered. However, the initial goal of these results is to check if the modeled system and the relevant protocols behave as expected. A method is based on the observation of sequence charts of the simulations generated when the event logging is turned on.

This type of result is also useful in cases of pedagogical objectives. Hence, Figure 4.4 shows the sequence charts of MAC and AVLC layers. As expected, when the channel is sensed free, the competing nodes send frames with probability *p* or wait for a *TM1* time slot with probability *1−p*. As soon as one node obtains the right to transmit, the frame is broadcasted and the other competing nodes will wait until the end of the transmission to continue the process. The AVLC protocol provides flow control and packet loss detection. The chart shows the use of the timers *T1* and *T2*. The first one helps to detect packet loss as when it expires, the sender retransmits the previously sent frames that are still unacknowledged. The timer T2 is used on the receiver side to slightly delay the acknowledgments in order to maximize the probability of sending it with eventual outgoing frames (piggybacking).

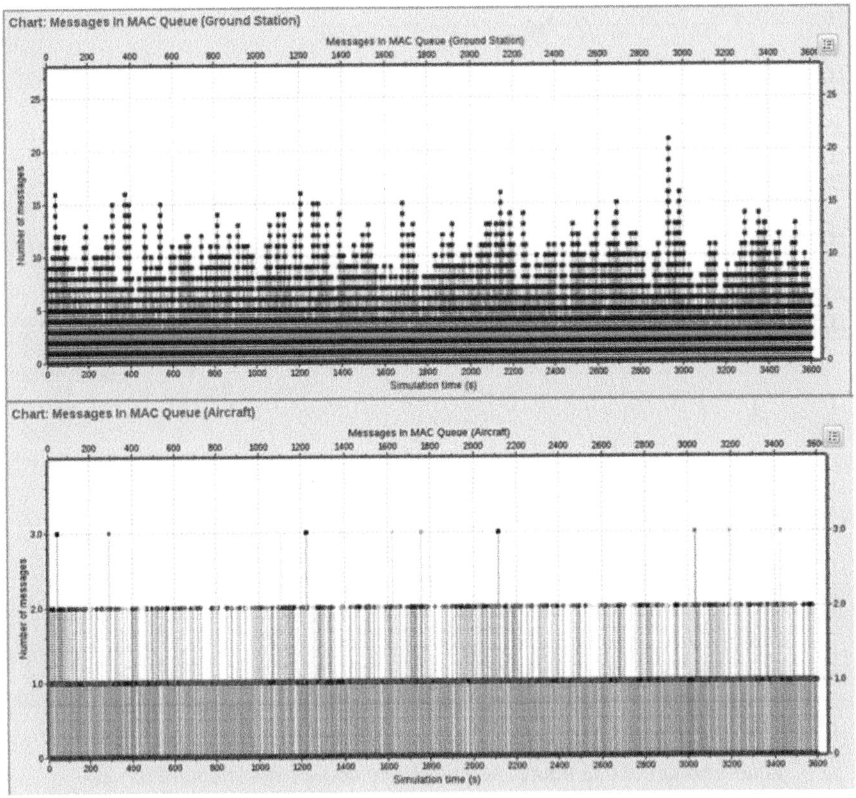

Figure 4.5. *Number of frames in MAC queue. For a color version of this figure, see www.iste.co.uk/hilt/transportation.zip*

Figure 4.5 helps to analyze the fill rate of the MAC queues in both aircraft and ground station. In this simulation, 100 aircraft are present in the VDL cell. In accordance with the results of existing studies and as previously explained, the generated messages in each application have a random length using a uniform distribution between 32 and 265 bytes. And the time between generated messages has a random value using an exponential distribution with a mean value of 40 s. The mean number of frames in the MAC queue is about seven times greater than that in the ground station. This is explained by the fact that each aircraft is connected to one ground station and the ground station is connected to several aircraft, 100 in the considered case. Hence, 100 DLEs feeding one MAC queue are present in the ground station.

Nevertheless, the results presented in Figure 4.6 show that the mean waiting time in the MAC queue is approximately similar for the ground station and the aircraft. This is explained by the fact that as the physical frames are quite long, relative to the MAC frames, their aggregation particularly benefits the ground station in the considered conditions.

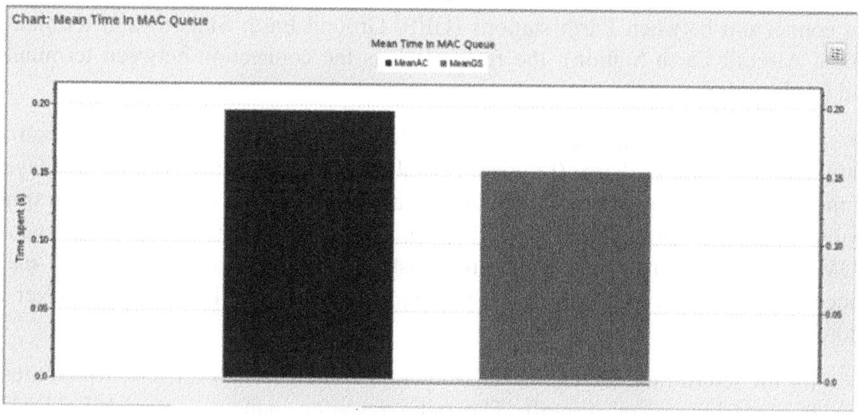

Figure 4.6. *Mean time in MAC queue. For a color version of this figure, see www.iste.co.uk/hilt/transportation.zip*

4.3. Oceanic air-ground data link communications and AMS(R)S

4.3.1. *The aeronautical mobile satellite (route) service and Classic Aero*

Telecommunication satellites play an essential role in the aviation context because of their ability to provide a worldwide service. Two types of system are currently certified by ICAO: the first, called "Classic Aero", is based on geostationary satellites (mainly the Inmarsat fleet) and the second is based on the IRIDIUM constellation of low Earth orbit (LEO) satellites. Both systems provide a data link supporting ACARS, FANS and ADS-C services and voice communications. They both provide low data rates when compared to those usually encountered in satellite communications (a few hundred of bits/s to a few kbit/s).

The architecture of the "Classic Aero" system is quite representative of that used for most communication systems employing geostationary transparent payload satellites (also called bent-pipe satellites). The salient features are:

– the satellite coverage is very extensive; splitting the service area into zones is necessary from both a performance standpoint (link budget) and frequency spatial reuse (total system capacity). In the case of "Classic Aero", coverage of one geostationary satellite corresponds to all of the visible area from a geometric viewpoint, about one-third of the Earth's surface excluding polar regions. This large service zone is then subdivided into regional beams (19 for one INMARSAT 4 satellite);

– topologies on forward and reverse links are different. The forward link is the connection between Earth stations (GES, Ground Earth Station) and terminals (AES, Aircraft Earth Station); the return link is the connection between terminals and Earth stations;

– for the forward link, the system takes advantage of the broadcast signals from a station to a geographical area (regional or global coverage beam). The physical layer of the system being one broadcast to all, the access method is quite naturally a time multiplex (TDM, Time Division Multiplex). An Earth station transmits on several TDM carriers continuously. Data broadcasted toward a group of aircraft may concern one or more of them; actual reception is based on filtering on the layer 2 address;

– for the return link, the available bandwidth is divided into carriers, which must then be used by several aircraft. The radio resource management is MF-TDMA (Multi-Frequency Time Division Multiple Access). Before sending data, each aircraft must identify both the appropriate carrier and time slot. One time slot accommodates a burst.

It is notable that access techniques in the Internet and multimedia geostationary satellites systems are designed on the same principles, even when the data rates are not comparable (several hundreds of Mbit/s). As an example, DVB-S2 implements TDM for the forward link and DVB-RCS2 implements MF-TDMA for the return link.

4.3.2. *Dimensioning parameters and bottlenecks*

Considering the architecture of "Classic Aero", the design and dimensioning of the forward link is rather straightforward. The capacity of a given carrier is set by the link budget; the needed number of carriers is determined by the total number of active aircraft within a beam. A simple queuing model allows for delay and congestion analysis. Conversely, the reverse link uses an access method, whose performance analysis can be tricky. MF-TDMA supposes that an aircraft identifies a time slot on a radio frequency carrier before sending one burst. Two access methods are implemented:

– a random access similar to S-ALOHA. This random access is of course used for network entry and the corresponding signaling but also for data transmission. The corresponding physical channel is called R for Random;

– a deterministic access. A signalization loop allows an aircraft to apply for a transmission capacity to the Earth station and obtain the allocation of a time interval on a carrier for data transmission. The corresponding physical channel is called T for TDMA.

The coexistence of these two access techniques for data transmission is justified by the delay induced by the geostationary satellite hop (about 250 ms). In the case of small data volumes, random access reduces the latency despite the lower efficiency. The downside is that as soon as the data volume to be transmitted becomes more consistent, the probability of data loss by collision and consequently the probability of retransmission may lead to degraded performances.

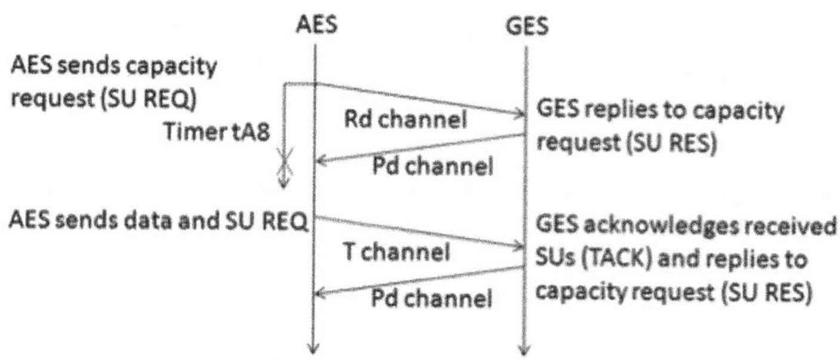

Figure 4.7. *R and T channels access procedure for Classic Aero return link*

The return link for "Classic Aero" relies on 11-byte data blocks (SU, Signaling Units) as a format for burst construction by the MAC sublayer. One block can be accommodated within one R channel burst. The MAC sublayer must determine whether one data block from the LLC sublayer should be transmitted over the random access physical channel R or using the deterministic access physical channel T. The decision is based on a simple threshold: the MAC sublayer switches to deterministic access as soon as the volume of data to be transmitted exceeds 33 bytes or three blocks. Figure 4.7 illustrates the signaling process for capacity requests and T slot allocation (note the retransmission timer tA8).

The main issues when designing the system are to ensure the random access technique runs in stable mode and to verify the T channel capacity is suited to that allocated to the R channel (distribution of carriers between the two physical channels). The main metrics to be investigated are:

– the random access channel R total load G;

– the random access channel R utilization S;

– the transmission delay for SU blocks over the R channel;

– the utilization of the T channel;

– the transmission delay as measured in the LLC sublayer.

The maximum delay observed in 95% of the cases is a system characterization driver.

4.3.3. *Simulation model*

The simulation model focuses on the analysis of the return link within one beam. Access to R and T channels is simulated with packets sent in radio bursts accommodated in each time slot. The information carried by the forward link is not broadcast but sent from point to point with a delay simulating the one induced by the satellite hop. The number of active aircraft is a simulation parameter; the ability of OMNeT++ to dynamically instantiate objects is thus exploited to change the network load. Figure 4.8 shows the appearance of the interface after loading the model.

The communications are point to point (aircraft to Earth station or opposite direction). The traffic model is similar to the one developed in the VDL Mode 2 model. And with a similar approach, modules of traffic generation and logical link management are instantiated in the Earth station at each entrance of an aircraft in the network. The model is a specific development and does not make use of model libraries like INET; for example, the addressing process relies on 3-byte aircraft identifiers as defined by ICAO (International Civil Aviation Organization).

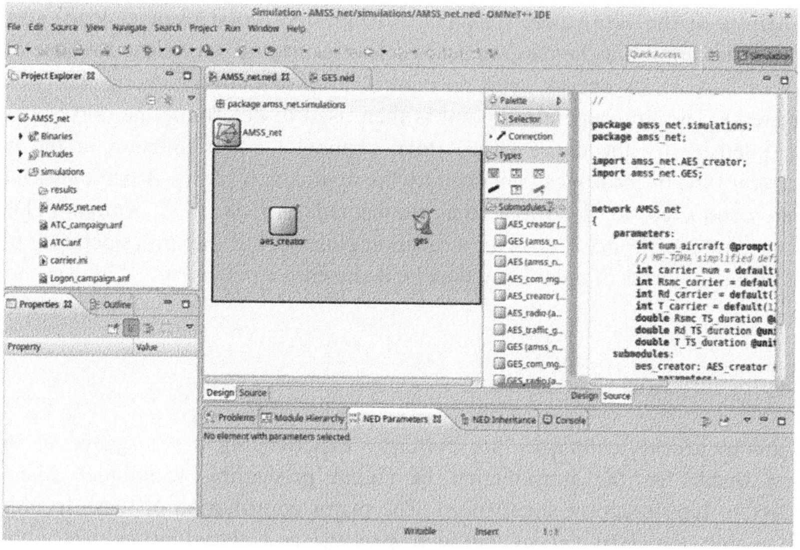

Figure 4.8. *AMSS simulation model. For a color version of this figure, see www.iste.co.uk/hilt/transportation.zip*

4.3.4. *Analysis of simulation results*

The primary objective of the simulation is to enable the analysis of the operation of the access layer and to establish a balance between the R and T channel capacities. The approach is shown in Figure 4.9.

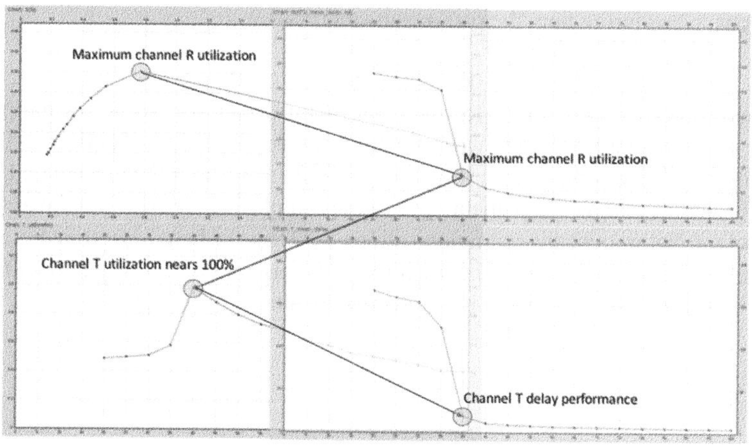

Figure 4.9. *R and T channel dimensioning*

Analysis of the S(G) trace (channel utilization vs. total channel load) sets the limit operating point of the random access. The results are very close to those of the theory of S-ALOHA access, with no control loop (e.g. by changing the back-off parameters). The limit operating point is then used to determine the corresponding traffic intensity by the curve G(iat) (total channel load in function of the mean interarrival time of messages generated by the application layer). A parametric study is then conducted to determine the number of carriers in T format (TDMA) necessary in order to get T channel utilization close to 1 at the limit operating point. Performance in terms of delay may then be deduced.

4.4. Summary and further work

Random access techniques are currently experiencing a resurgence of major interest thanks to the introduction of signal processing techniques like SIC (Successive Interference Cancellation). The major contribution of these techniques is to enable the retrieval of collided packets and therefore greatly improve performance. In the context of aeronautical satellite communications, the proposed standard enacted as part of the IRIS project [IRI 13] is based on the use of the E-SSA access method (Enhanced Spread Spectrum-Aloha). OMNeT++ is a very suitable tool for studying the performance of such systems, in which the characteristics of the mobile radio channel have a significant impact (distribution of signal powers at receiver input in particular).

Furthermore, random access techniques may also be driven by several parameters similarly to CSMA p-persistent in VDL mode 2. And here again, OMNeT++ is very useful to study the performance of the system under different conditions, considering how the parameters are tuned.

The presented models can of course be improved, particularly by including simulated aircraft trajectories. However, the main driver for further developments will be to build a unified framework for the considered systems (VHF and Satcom data links) and new architectures in order to be able to address the present and future research challenges. We can mention as an example the vertical handover in the presence of heterogeneous communications systems, where on-ground network interconnection and aircraft on-board router designs interact in order to fulfill the reliability, availability and delay performance objectives of ICAO.

4.5. Bibliography

[BEN 13] BEN MAHMOUD M.S., GUERBER C., LARRIEU N. *et al.*, *Aeronautical Air-Ground Data Link Communications*, ISTE and John Wiley & Sons, 2014.

[ICA 07] ICAO, "Annex 10 to the Convention on International Civil Aviation, Volume III Communication Systems (Part I Digital Data Communication Systems, Part II Voice Communication Systems)", 2007.

[IRI 13] IRIS, "ANTARES Communication Standard Technical Specifications", IRIS-AN-CP-TNO-612-ESA-C1, Issue 1.0, September 2013.

[OMN 16] *OMNeT++ Discrete Event Simulator*, available at: https://omnetpp.org/, 2016.

[VAR 08] VARGA A., "An overview of the OMNeT++ simulation environment", *Simutools '08 Proceedings of the 1st International Conference on Simulation Tools and Techniques for Communications, Networks and Systems & Workshops*, p. 60, 2008.

5

A Virtual Laboratory as an Assessment Tool for Wireless Technologies in Railway Systems

The European Rail Traffic Management System (ERTMS) is now an international reference standard for railway signaling. Its deployment in Europe will be long and expensive; thus, the industry needs faster rollout and a reduction in cost in order to obtain the certification and authorization to put equipment into service. Most of the proposed lab-testing tools for ERTMS evaluation have focused mainly on its functional subsystem, the European Train Control System (ETCS). The related test scenarios and simulators have been developed while assuming an ideal telecommunication subsystem. On the contrary, wireless technologies suitable for supporting communications in railway systems have been evaluated only on the basis of the key performance indicators expected for ERTMS signaling applications. Most of the evaluations in the literature have considered only the performance evaluation of the wireless technology itself, using a network simulator, and without taking into account effective train traffic scenarios and the related ETCS feedback. This chapter presents a virtual laboratory based on co-simulation and relying on two existing tools: an ERTMS simulator implementing the functional subsystem (ETCS) and an OPNET simulator that allows us to model the whole telecommunication subsystem, namely the GSM-R (Global System for Mobile Communications – Railways). First, the virtual laboratory architecture and the assumptions on which it has been built are presented. Then, this chapter describes how offline and live co-simulation between the aforementioned simulators can be performed. Thus, the impairments of any prospective wireless technology can be taken into account during the simulation-based evaluation of ERTMS traffic scenarios before costly real-world testings. Finally, the virtual laboratory serves as a case study in the analysis of the co-simulation approach, particularly when the simulators are not of the same type. Prospective work targeting an evolution from co-simulation to multi-modeling, in order to directly connect the models and avoid the problems related to heterogeneity of simulators, concludes the chapter.

Chapter written by Patrick SONDI, Eric RAMAT and Marion BERBINEAU.

5.1. Introduction

The International Union of Railway (UIC) introduced the European Rail Traffic Management System (ERTMS – visit www.ertms.net) in order to harmonize the different train control systems in use in Europe. Following the goal of opening the market of customers and goods transportation inside the European Union (UE), this harmonization also needed to be accompanied by an optimized utilization of the tracks through dynamic train control. In order to achieve this objective, the ERTMS needed both safer train driving supervision processes and a continuous train-to-track/track-to-train communication system able to operate at high-speed levels. These two functions are mainly ensured by the two major components of the ERTMS: (1) the telecommunication subsystem, GSM-R (Global System for Mobile Communication – Railway), which ensures wireless communications between the train and the control location and (2) the functional subsystem identified as the European Train Control System (ETCS), which ensures control of the train and its signaling with the control location via the GSM-R infrastructure [RUE 08].

As a set of control–command processes, the ETCS applications are prone to evaluation approaches that only need to prove their correctness, such as formal methods. The Union Industry of Signaling (UNISIG), the consortium in charge of the development of ERTMS/ETCS technical specifications, has produced the subset 026 [UNI 10] that fixes the compliancy requirements for any test bed dedicated to ERTMS evaluation. Some ERTMS simulators are presented in [MER 07], and the one used in this work is compliant with the subset 026. However, although the functional behavior of the components communicating through the GSM-R infrastructure, such as the Radio Block Center (RBC), is modeled in these simulators, the underlying telecommunication technology itself is not modeled. As a result, GSM-R communications are supposed to be ideal and their related failures due to the impairments of wireless communications, to mention a few, cannot be taken into account during the evaluations performed with these simulators.

On the contrary, it should be noted that GSM was almost the most widely deployed wireless mobile technology in Europe when the telecommunication subsystem for the ERTMS needed to be specified. Thus, the main question was not "which mobile technology for the ERTMS?", but almost "can GSM do the job?". It took several years of real-world testing to demonstrate that the GSM-R satisfies the Quality-of-Service (QoS) requirements imposed by ERTMS/ETCS applications at the telecommunication subsystem interface. Nowadays, GSM is a declining technology, and several technologies widely deployed over the world could be potential solutions to replace it in the ERTMS. Moreover, nowadays nobody can imagine performing several years of real-world testing for each one of these prospective solutions. In this context, the use of virtual laboratory could evaluate them by simulation and determine how they meet the QoS requirements of current

ERTMS/ETCS applications and which new value services they could introduce while maintaining or even improving safety and security.

The literature contains some attempts at modeling ERTMS/ETCS applications directly inside network simulators such as OPNET in order to include the model of the telecommunication subsystem technology during the evaluations [RUE 08, LOP 14]. However, these evaluations are very limited because of the difficulty of modeling every ETCS application while also including all the ERTMS components and all the factors that may affect an ERTMS traffic scenario. For these reasons, this work proposes a co-simulation approach, where an ERTMS simulator can be connected to a network simulator that models the telecommunication subsystem. In this way, the resulting virtual laboratory can rely on an ERTMS evaluation tool compliant with the subset 026, while taking into account realistic behavior of the telecommunication subsystem.

The presentation of this chapter is organized as follows. Section 5.2 presents the main features of both the functional and the telecommunication subsystems of the ERTMS and some work related to their evaluation by simulation. The co-simulation approach developed in this work, the resulting virtual laboratory architecture and implementation are presented in section 5.3. A case study of an ERTMS scenario evaluated using this virtual laboratory is presented in section 5.4, completed by a discussion on the advantages and drawbacks of the co-simulation approach itself. This chapter is concluded in section 5.5, where prospective works are also announced.

5.2. ERTMS subsystems and related test beds

The different components of an ERTMS deployment are described in [MID 08], the main two being the functional subsystem and the telecommunication subsystem.

5.2.1. *The functional subsystem of the ERTMS*

This first component is the European Train Control System (ETCS), which is dedicated to train signaling and control. It is designed in order to fulfill the following three main objectives [LEV 08]:

– Improved safety by train driving supervision: during its movement, the train receives information about running limitations (speed, distance, etc.) in the form of a Movement Authority (MA) that defines a place on the track (End of Authority – EO

As A), which it must not pass. On the basis of both the track and train data, the on-board equipment calculates a set of braking curves for train movement supervision;

– Higher performance by increasing speed and capacity: provided the movement information directly through displays, the driver can drive safely following the speed limitation until the next EOA without having to look at trackside signals;

– Interoperability: in contrast to trackside signaling systems based on colors depending on national rules, ETCS is an appropriate train control system for lines belonging to different railway administrations.

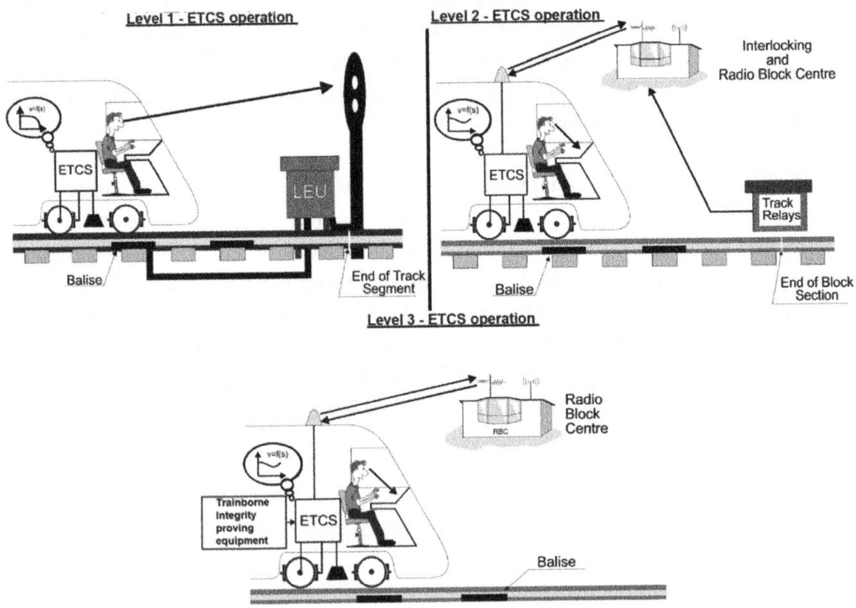

Figure 5.1. *ETCS operational levels (image from [LEV 08])*

To achieve these goals progressively on the different railroads, ETCS specifications define different ETCS implementation levels for lines in relation with trackside equipment (Figure 5.1). In ETCS level 2, a train equipped with ETCS operates on a line controlled by a Radio Block Center (RBC) and is equipped with Eurobalises and GSM-R. The train is permanently connected to the RBC using GSM-R infrastructure. In this way, the control center can update the information about train movements in real time and supervise them more dynamically. Current ERTMS deployments involving GSM-R concern this ETCS level.

Several research works targeting ETCS level 3 are still in progress, especially concerning the use of a satellite-based localization system in railway transport [BEU 12]. Thus, all references to ETCS in this chapter implicitly concern level 2.

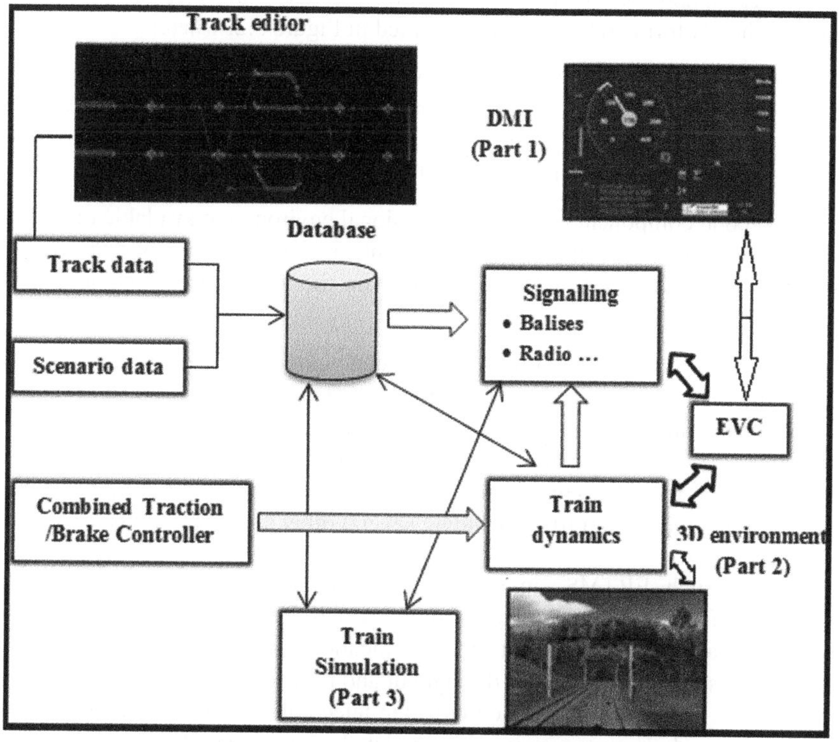

Figure 5.2. *ERTMS simulator architecture (image from ERSA France)*

The ETCS applications play a key role in the safety and efficient supervision of railway traffic. For this reason, their conception and evolution follow a stringent validation process. In such a process, test beds are particularly useful in order to perform fast and low-cost preliminary evaluations. Almost all existing ERTMS simulators are designed in order to only evaluate the functional behavior of the system. It is possible to verify, in normal functioning conditions, if a control–command procedure makes the train behave as expected in the specifications. The proposed ERTMS simulators are validated on the basis of the fact that they allow all the tests required in the subset 026 to be performed [UNI 10]. The ERTMS simulator used in this work has been implemented following the subset 026 specifications, and the resulting simulation platform is compliant to the requirements for ERTMS test beds. It consists of three main components and several additional offline tools for scenario design and analysis distributed over computers connected through a wired network. The platform architecture is described in Figure 5.2, where:

– a train driving simulator equipped with a Driver Machine Interface (DMI) compliant to CENELEC specifications is attached to the first component. A human operator can virtually drive the train on a single ERTMS track. The data of the scenario are stored for post-simulation analysis;

– the second component is a three-dimensional environment available on a single track. When used with the first component, it reproduces a virtual realistic environment for the driver, who can also rely on the visual signals included in the scenario;

– the third component consists of several modules: a route manager, an interlocking management system, including up to two RBCs, up to 11 trains moving simultaneously, and also the driving simulator. This component is both the control center of the railway traffic and the trains' manager in manual or automatic mode. When used with the other components, it allows the human operator on the driving simulator to interact with traffic, including several other simulated trains.

Although such ERTMS simulators usually include a GSM-R interface, the functioning of the telecommunication subsystem is idealized. Therefore, it is not possible to evaluate the values of telecommunication-related metrics such as end-to-end delay, loss rate, network load, throughput and retransmission count per message. Moreover, the impact of a dysfunction in the telecommunication subsystem on the behavior of the whole system cannot be simulated with these tools.

5.2.2. *The telecommunication subsystem of the ERTMS*

The telecommunication subsystem is actually the second major component of the ERTMS. The main part of the ERTMS level 2 is currently implemented using the GSM-R (Figure 5.3). This technology is based on the classical GSM architecture, but it uses specific frequency bands dedicated to railway communications. In France, frequency bands from 876 to 880 MHz are used for uplink transmission (Mobile Station – MS – to Base Transceiver Station – BTS), and those from 921 to 925 MHz are used for downlink transmission (BTS to MS). There are 20 channels of 200 kHz each uplink and the same amount for downlink to allocate to the different BTSs, which are placed every 3–4 km along the railway in order to ensure high redundancy and to support high speeds up to 500 km/h.

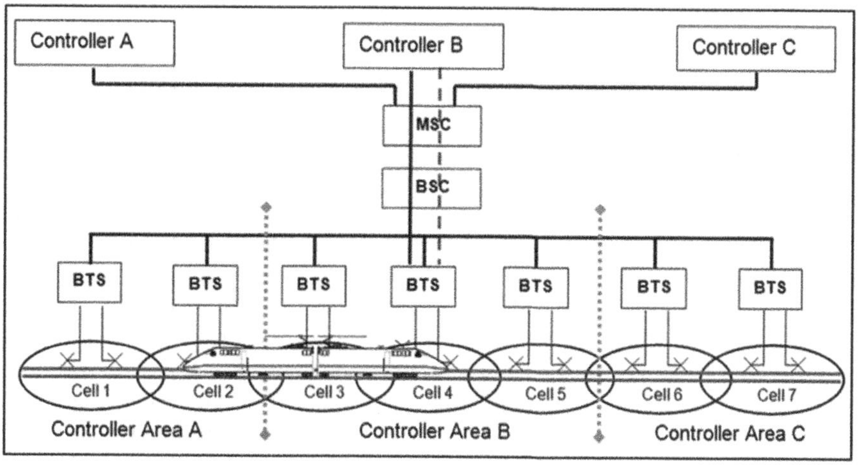

Figure 5.3. *GSM-R infrastructure with redundant BTS (image from Siemens)*

The telecommunication subsystem plays a key role in the ERTMS as it ensures communications between the control center and the train for the traffic related to both the signaling and the applications. For these reasons, stringent requirements have been specified for the telecommunication technologies candidate that could be adopted in ERTMS. GSM technology met these QoS requirements, and the tests carried out by the European Railway Agency (ERA) confirmed its accuracy as a telecommunication technology for the ERTMS. GSM technology also has two other major advantages as a solution for interconnecting railway networks of the different European countries:

– GSM was widely deployed by mobile telephony operators in Europe, and for this reason, both the equipment and maintenance costs were relatively lower than those of other technologies;

– in addition to ETCS signaling, GSM was used by almost all European railroad operators for their professional mobile communications. In this way, the same technology served to interconnect both the ERTMS infrastructures and the operators' professional networks of the different countries.

Since the adoption of GSM-R, many developments have occurred in transportation. Indeed, the development of Intelligent Transport Systems (ITS) has brought new applications for the safety and monitoring of transport systems and also new services for customers and user-friendly applications. Evolution of railway transportation systems in order to provide some of these new services will be mandatory for its competitiveness; however, it will also imply new QoS constraints and consequently increase the traffic supported by the communication network. In this context, the

GSM-R may still not be the appropriate technology [SON 12]. Several researchers proposed investigating other telecommunication technologies for the ERTMS, such as GPRS [RUE 08], WIMAX [AGU 07] and, recently, LTE [SNI 14]. Analytical and simulation-based evaluations on these telecommunication technologies are proposed in the literature regarding various telecommunication-specific metrics.

The authors of the aforementioned work used the Riverbed OPNET modeler, which is one of the most popular simulators for the evaluation of network technologies. However, their experiments concerned only the behavior of the telecommunication subsystem and were disconnected from the functional part of the ERTMS. The ETCS applications evaluated are modeled approximately in terms of the messages that they generate during the simulation scenarios; however, the behavior of the functional component of ERTMS is not actually modeled in these scenarios. Consequently, it is possible to evaluate the value of the telecommunication-specific metrics for some particular messages exchanged during the scenario, whereas it is not possible to actually observe the behavior of ETCS applications in a specific ERTMS scenario when a dysfunction occurs in the simulation of telecommunication technology.

These observations emphasize the need for an evaluation tool for the ERTMS in which both the functional and telecommunication subsystems can be simulated and in which the impact of the behavior of one component on the functioning of the other component can be studied accurately. It is the purpose of the work presented in this chapter, where OPNET is also used as a telecommunication simulator.

5.3. A virtual laboratory based on co-simulation for ERTMS evaluation

This section presents the co-simulation approach developed in the ANR Project VEGAS (**V**irtual lab based on co-simulation to include impairments of wireless t**E**lecommunication such as **GS**M-R in the ev**A**luation of ERTM**S** components) that supported this work and the resulting virtual laboratory software tool.

5.3.1. *Why a co-simulation approach?*

As mentioned in section 5.2, current ERTMS simulators are mostly designed to evaluate the functional behavior of the system, and they have been validated for this purpose. However, they do not actually implement the telecommunication subsystem and do not allow for evaluation of either the behavior of the entire system regarding telecommunication metrics or the scenarios where dysfunctions occur in the telecommunication subsystem. Integrating a model of the telecommunication subsystem inside the current ERTMS simulators would require a complete development, from scratch, of all the components of the GSM-R architecture and from the physical to the network and transport layers. Moreover, it would be necessary to develop the models for all other prospective telecommunication technologies and maintain the evolution of the related protocols and equipment in the designed models. Such work would be equivalent to that of designing a complete network simulator from scratch, and it should be avoided as efficient and validated tools, such as OPNET, already propose powerful features for advanced simulation of network and telecommunication technologies.

We also noted that the functional subsystem of ERMTS is made of various ETCS applications. Implementing all the features of these applications in a network simulator would result in an inefficient modeling of the complete ERTMS functional subsystem again, as it is already done in current validated ERTMS simulators. Moreover, keeping this validation for the resulting platform would not be straightforward.

To avoid such complicated and unpredictable work, we propose a new approach based on co-simulation that will connect an ERTMS simulator with a simulator especially designed for network and telecommunication technologies, namely the OPNET simulator, in order to design a simulation tool dedicated to the joint evaluation of the functional and telecommunication components of ERTMS.

5.3.2. *Which data and processes must be modeled in each simulator?*

In any ERTMS level 2 scenario, each train moves on a specific track as described in Figure 5.1. On its movement through GSM-R, the train sends various information to the control center via the RBC and receives specific instructions (MA) in the same way. Therefore, under the assumption that all the communications occurring in the scenario between the train and the RBC meet the requirements imposed by ERTMS at the GSM-R interface, the behavior of the functional subsystem can be accurately evaluated using the ERTMS simulator.

Following the same reasoning, let us consider a scenario simulated on the ERTMS simulator, where the movement (successive positions, instant velocities and accelerations, etc.) of the train over a certain time as well as all the sequence of the messages exchanged in that time with the RBC during this movement are completely stored. Under the assumption that we are able to precisely reproduce the same movement in OPNET and the same sequence of messages following the same chronology, it is possible to precisely obtain the value of the end-to-end delay for each message exchanged. Other telecommunication-related metrics can be studied in the same way.

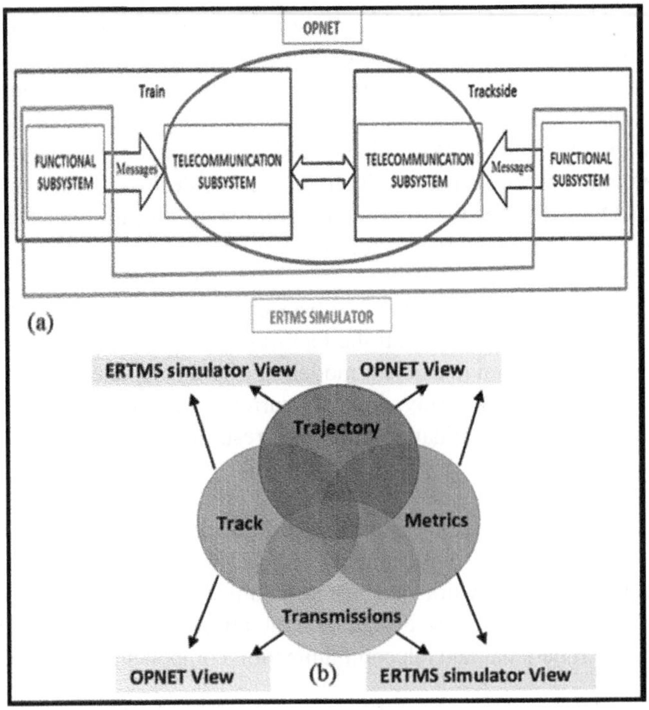

Figure 5.4. *Co-simulation architecture and concepts. For a color version of this figure, see www.iste.co.uk/hilt/transportation.zip*

Following these ideas, the co-simulation platform architecture can be described as in Figure 5.4(a). The functional subsystem is simulated using the ERTMS simulator, and the telecommunication subsystem is simulated using OPNET. The key elements that synchronize both simulators are the movement of the train and the messages exchanged with the control center during this movement.

Therefore, the same ERTMS scenario can be partially modeled inside the different simulators in order to evaluate the related components. To ensure the coherency of each one of the partial scenarios modeled inside each simulator with the ERTMS scenario, the following concepts are introduced [SON 12]:

– The Track: this concept represents the physical and static elements that materialize the railroad, the network infrastructure, the localization and signaling systems component;

– The Trajectory: this refers to the movement of one train during a specific scenario. In this way, the movement of any train during an ERTMS scenario can be reproduced faithfully inside any of the simulators;

– The Transmissions: these refer to the set of messages exchanged between each train and the control center during a scenario, ordered by the date of emission;

– The Metrics: these refer to the indicators that are evaluated during a scenario. In an ERTMS functional simulator, we can mainly evaluate the conformity of the train behavior with the ERTMS safety specifications. In a telecommunication simulator, we can evaluate metrics such as end-to-end delays, loss rate and handover duration.

When generating the partial view of an ERTMS scenario for a specific simulator, a specific view must be generated for each one of these four concepts (Figure 5.4(b)). The related view to generate would contain more or less details, according to the component evaluated by one specific simulator and the related metrics considered.

5.3.3. *Overall architecture of the ERTMS–OPNET virtual laboratory*

The virtual laboratory is realized through a software infrastructure that connects the ERTMS simulator and OPNET. It is composed of the following three major components (Figure 5.5):

– The ERTMS co-simulation interface: this proposes a set of remote procedures that can be invoked in order to obtain either trajectory information about the trains or the messages emitted by both the trains and the RBCs involved in a scenario. These interfaces are proposed as CORBA interfaces: the RMCPlugin (over the Route MaP Controller) for trajectory information and the RNSPlugin (over the Radio Network Simulator) for the messages;

– The ESYS interface: each node model in OPNET (train or RBC) contains an ESYS process, which exposes an interface able to exchange data with an external program. In this way, each train model in OPNET can be notified of any change in the position of the corresponding train in the ERTMS simulator so that it can update itself. Also, any message sent by a train or RBC in the ERTMS simulator is written at the interface of the corresponding train or RBC model in OPNET so that this latter performs the emission of the message in the OPNET as well;

– The co-simulation manager: it is composed of two independent components that share some data, namely the movement manager and the message manager. The first connects to the RMCPlugin and obtains information about the scenario, the track, the trains and the RBCs. Periodically, it requests current train position in the ERTMS simulator to the RMCPlugin and sets them on the ESYS interface of the corresponding train in OPNET. Also, the message manager is notified by the RNSPlugin of each emitted message in the ERTMS simulator so that it can notify the

OPNET model of the corresponding emitter through its ESYS interface. When a node model in OPNET receives an ETCS message, it sends feedback to the message manager through a callback function registered on its ESYS interface so that the latter can notify the RNSPlugin.

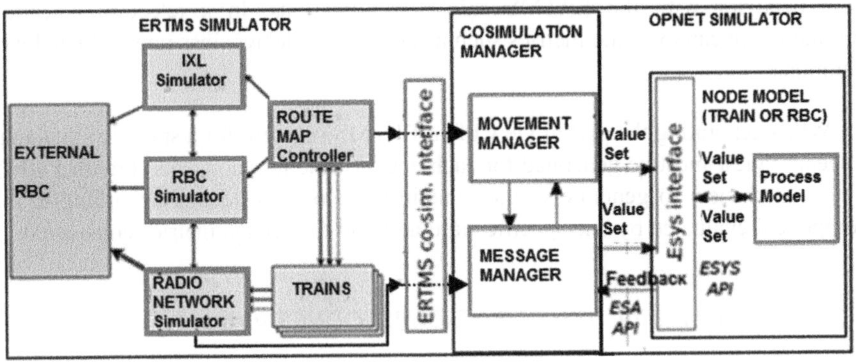

Figure 5.5. *ERTMS–OPNET virtual laboratory architecture*

5.3.4. *Synchronization modes*

In order to perform a co-simulation, both the ERTMS simulator and OPNET can be connected live for an online simulation or offline by replaying a scenario previously run in one simulator in the other.

Online simulation presents the advantage of running both functional and telecommunication subsystems simultaneously, thus allowing a more realistic evaluation of the entire ERTMS scenario. In this mode, the model of the telecommunication subsystem implemented in OPNET interacts with the Radio Network Simulator and the Route Map Controller of the ERTMS simulator (Figure 5.5). The trajectory information of the train is transmitted live to OPNET so that the train follows the same movement in both simulators. The messages generated, for example, by a train in the ERTMS simulator go without delay through the OPNET train model, where they are sent to the OPNET RBC model through the GSM-R infrastructure modeled in OPNET under realistic conditions of mobility, propagation and network transmissions. They are then routed without delay from the OPNET RBC model to the RBC simulator in the ERTMS simulator for functional processing. The complete process followed by one message and the related response during co-simulation are illustrated in Figure 5.6.

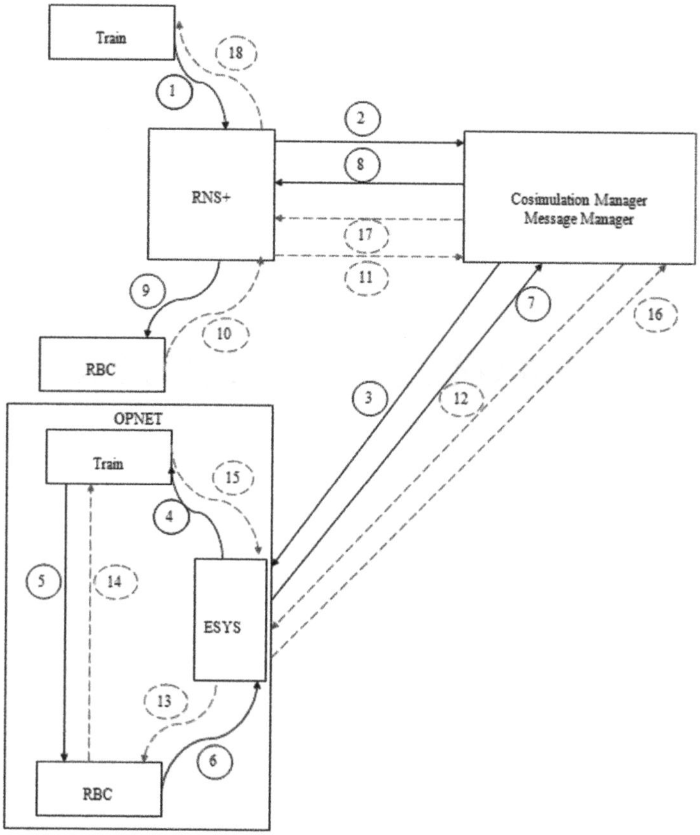

Figure 5.6. *Process of a message and its related response during online co-simulation*

For any scenario run with the ERTMS simulator, the information about the track, trajectory, transmissions and metrics is backed up as chronologic events by the co-simulation manager. In this way, it is possible to check that the scenario is valid from a functional viewpoint. This scenario is then replayed offline in OPNET in order to evaluate the behavior of the telecommunication subsystem under realistic functional constraints. Typically, there is no difference between online and offline simulation at the OPNET interface. As a result, it is also possible to use real-world traces of the trains running on ERTMS level 2 lines in order to build offline co-simulation scenarios, provided that these traces are formatted like the data backed up by the co-simulation manager during online co-simulation [SON 13].

5.3.5. *Virtual laboratory implementations in the ERTMS simulator*

The implementation of the virtual laboratory in the ERTMS simulator is achieved through two interfaces: the RMCPlugin and the RNSPlugin.

The RMCPlugin implements remote procedures that allow any client to post updated information periodically or on-demand (Figure 5.7) on the movement of the trains in a scenario. The movement manager module of the co-simulation manager registers to the RMCPlugin server, and it sets the periodicity for receiving push messages about train position updates.

Information	Description	Relevant database field
Train identifier	Unique train identifier[1], value of the table primary key	*Train.TrainID*
Train name	Name of the train as entered in the Scenario Editor	*Train.Name*
ETCS identity	ETCS identity of the train	*EurocabParameters. ETCSIdentity*
Train running number	Running number of the train	originally *EurocabStartingConditions. TrainRunningNumber*[3]
Current train segment identifier	Identifier of the segment on which the train head is currenly located	originally *Segment.SegmentID*[3]
Current train segment offset	Offset (in meters) on the segment on which the train head is currently located	originally *Train.Position*[3]
Train coordinates	(X,Y) graphical coordinates of the train. See also 0	/
Train position	Offset (in meters) of the train head on the segment on which the train head is located	/
Train speed	Speed of the train (in km/h)	/
Train acceleration	Acceleration of the train (in m/s²)	/

Figure 5.7. *Movement information provided by the RMCPlugin of ERTMS simulator*

The RNSPlugin implements remote procedures that allow any client to register in order to receive a copy of any message emitted by a train or RBC in the ERTMS simulator. It is also able to receive a notification from the message manager of the co-simulation manager about the acceptance or the rejection of a message after its transmission has been simulated with OPNET. The functioning of the RNSPlugin server operating at the ERTMS co-simulation interface is summarized in the chart presented in Figure 5.8.

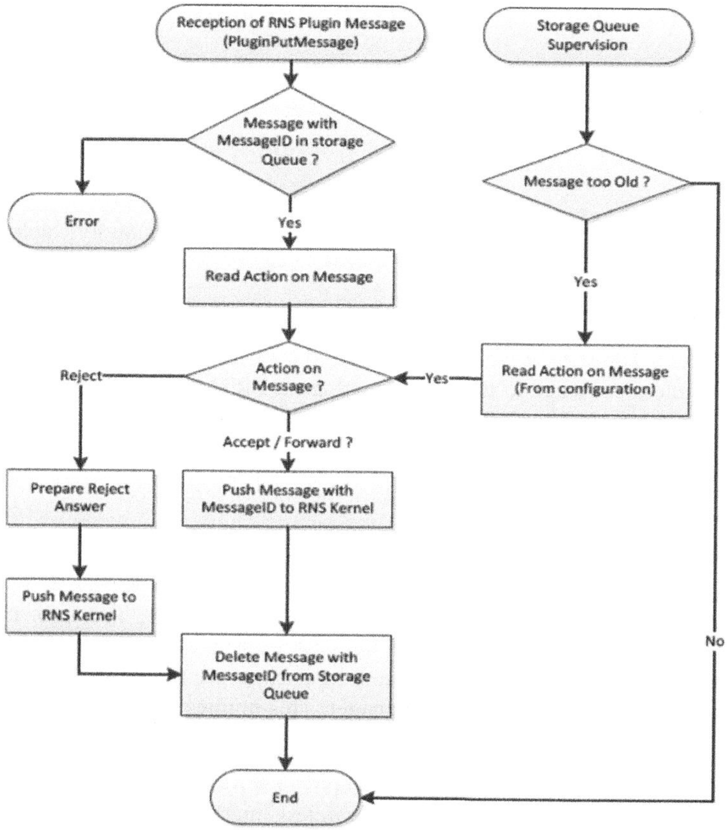

Figure 5.8. *Functioning of the RNSPlugin server at ERTMS co-simulation interface*

5.3.6. *Virtual laboratory implementations in OPNET*

In a classical GSM infrastructure, many components are involved in order to ensure efficient management of the network and provide the best services to the subscribers. However, a GSM-R infrastructure is dedicated only to ERTMS operations, and in this project, the focus is on the wireless impairments. As a result, the core network from the Base Transceiver Stations (BTSs) can be simplified as all the components are connected through a wired infrastructure.

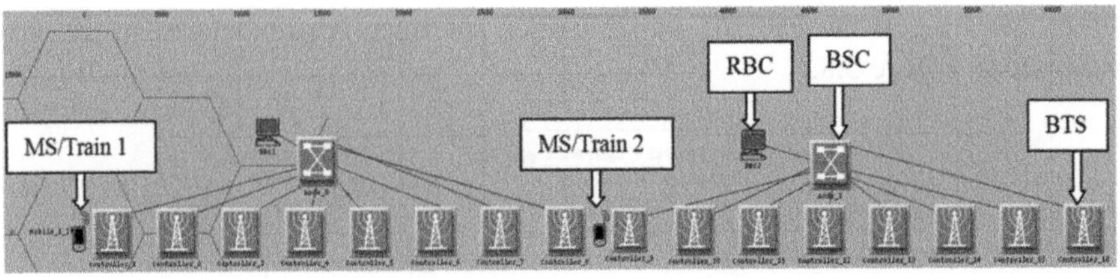

Figure 5.9. *GSM-R infrastructure in OPNET for co-simulation*

The GSM-R infrastructure modeled in the virtual laboratory includes the following components (Figure 5.9):

– the mobile nodes that model the trains;

– the Base Transceiver Stations that are placed along the railroad in order to provide wireless coverage to the trains running on ERTMS level 2 lines. Following GSM-R deployments, they are placed every 4 or 7 km depending on the railroad configuration;

– the Radio Block Centers that represent the control locations.

The train model is derived from an advanced wireless node model in OPNET. As a result, it contains all the components, making a refined modeling of all the protocols from the application to the physical layer possible. The train model in OPNET does not generate traffic directly. The messages are produced by ETCS applications in the ERTMS simulator, and the packets are routed to OPNET via the ESYS interface by the co-simulation manager. This implies the following:

– the ETCS applications do not need to be modeled in OPNET. Only the message that they generate in the train's on-board equipment is routed through OPNET in order to simulate the related packets through the wireless interface to the core network. The main advantage is that of preventing the very long development of only approximated models of the actual ETCS application with relatively poor gain in terms of ERTMS analysis. Another advantage is that any new ETCS application introduced in the ERTMS simulator will be able to be simulated immediately through this co-simulation architecture without any change to the train model;

– the messages coming from the ERTMS simulator include connection requests and releases. Furthermore, they are already managed by the EURORADIO layer modeled in the ERTMS simulator. As a result, there is no need to model this layer in OPNET. The ETCS messages are already encrypted, and any request related to a message not routed by OPNET due to wireless impairment will not have any effect. As a result, the connection or disconnection or data transmission processes and their implications are still managed by the ETCS application implemented in the ERTMS simulator, and only the consequences due to a wireless impairment are actually managed in OPNET.

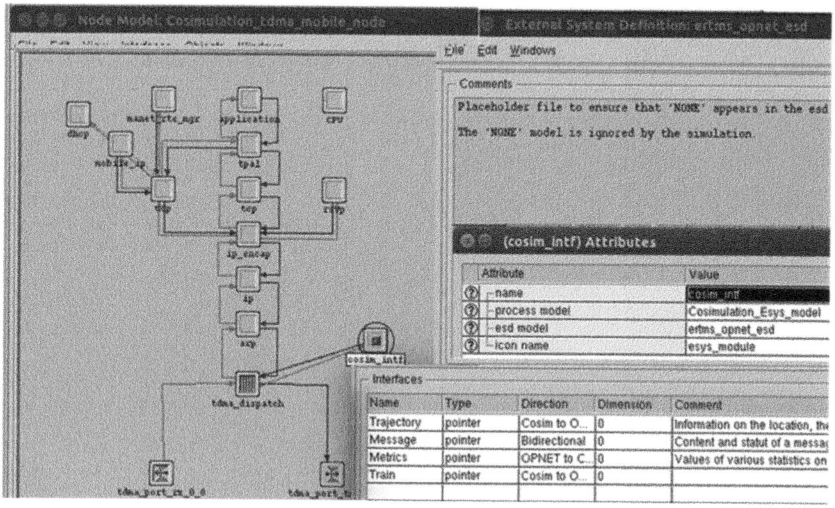

Figure 5.10. *Train model and its ESYS interfaces (cosim_intf) in OPNET. For a color version of this figure, see www.iste.co.uk/hilt/transportation.zip*

The BTS model is derived from the OPNET native model of a bridge node with two interfaces: a TDMA interface and an Ethernet interface. The TDMA interface manages wireless communications with the trains evolving in the location covered by the BTS, and the Ethernet interface connects each BTS to a central switch at the control location. In the global architecture, the group of BTS that depends on the same RBC is connected to the switch connected with this RBC.

The model of the RBC is almost the same as that of the train, except that there is no TDMA interface but an Ethernet interface instead. It can be noted that the same ESYS model is used for both the train and the RBC, following the generic approach developed in the VEGAS project.

5.3.7. *Virtual laboratory implementations in the co-simulation manager*

The co-simulation manager main window presents the main information about the scenario and the track as well as the state of its different components (Figure 5.11), namely:

– Train Manager: its state (running, stopped or not working) and position update periodicity;

– Message Manager: its state and the simulation time in ERTMS simulator;

– OPNET interface: its state and the current simulation time in OPNET.

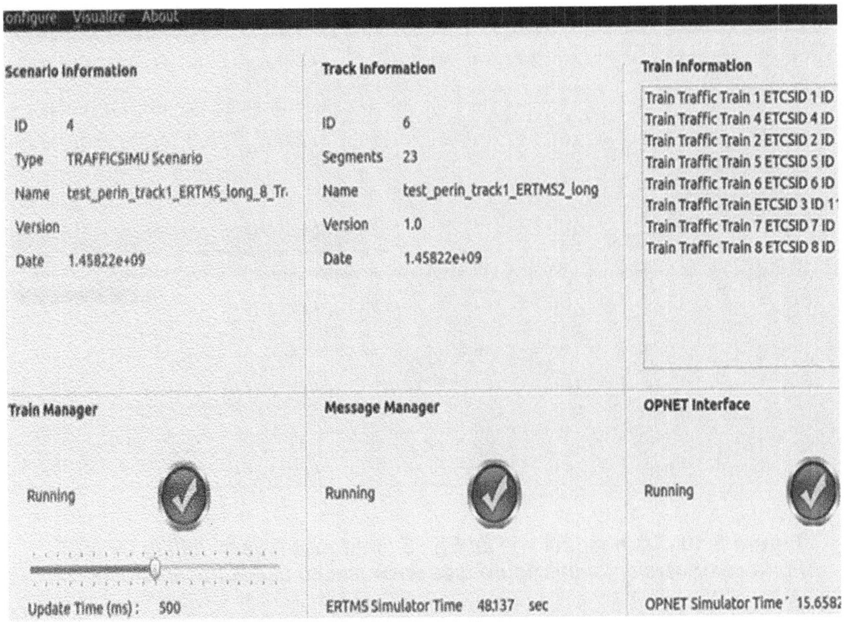

Figure 5.11. *Co-simulation manager main window*

During co-simulation, it is possible to monitor the train positions through the Visualize menu. Also, through another submenu of the Visualize menu, it is possible to monitor the messages sent or received by the different trains and RBC (Figure 5.12).

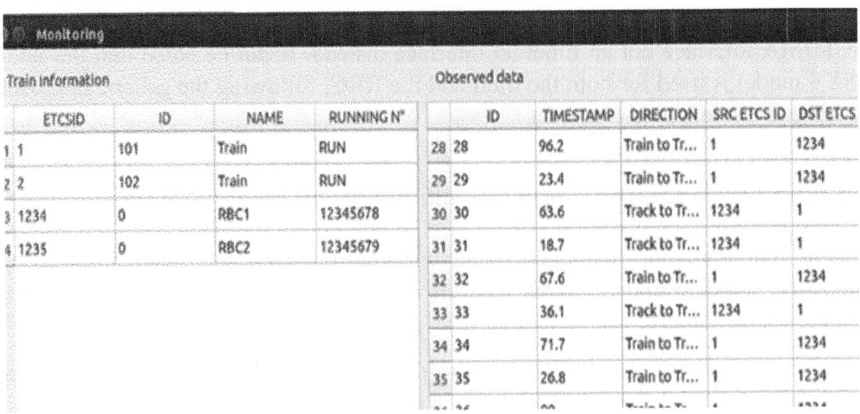

Figure 5.12. *Message view in visualize menu of the co-simulation manager*

5.4. Effective use of the ERTMS–OPNET virtual laboratory

5.4.1. *A co-simulation scenario with the ERTMS–OPNET virtual laboratory*

This section presents the co-simulation process applied to a scenario and the related main steps. While simulation is running, both the OPNET interface in the co-simulation manager and the OPNET simulator print debug traces in a terminal or a file. These traces include the following:

– OPNET model of the network scenario invoked in the co-simulation (Figure 5.13);

– ESA initializations, including the ESYS interfaces of predefined trains and RBC (Figure 5.13);

– indexes of the available interfaces for assignment to ERTMS simulator trains and RBCs; in this way, it is possible to perform the assignments automatically (Figure 5.13);

– the time in each simulator and the difference of time between them; in this way, it is always possible to know if the co-simulation can continue online or if it should go offline when synchronization conditions cannot be met anymore (Figure 5.13);

– co-simulation events: for example, when the OPNET interface receives train information from the co-simulation manager for Train with ETCSID 1 (Figure 5.14);

– interfaces assignment: for example, Train 1 is associated with ESYS interface 0 (Figure 5.14);

– OPNET traces always start with this label, for example, when nodes are initialized (Figure 5.14);

– replacement of default positions of nodes in OPNET by their initial positions in ERTMS simulator when trajectory information is received, and association between ETCSID of the train or RBC and their MAC address in OPNET (Figure 5.15);

– update of train position in OPNET upon receipt of new positions of the corresponding train in the ERTMS simulator through trajectory data (Figure 5.16). It can be noted that when no message is exchanged, the time difference between the simulators can be higher without causing synchronization problems (e.g. up to 0.5 s in Figure 5.16);

– message transmission: for example, Figure 5.17 shows that the message of ID 1 sent by Train with ETCSID 1, which is associated with Mobile_1_1 in OPNET, is sent to its current controller (Controller_1). The latter then sends it to the switch node_0, which transmits it to RBC1;

– when the message reaches the destination in OPNET (namely RBC1), RBC1 updates routing table information and sends a notification to the co-simulation manager to indicate whether or not the message should be routed in ERTMS simulator (Figure 5.18);

– some metrics values are available immediately during co-simulation (Figure 5.19); however, the detailed telecommunication metrics, such as end-to-end delays, loss rates, throughput, network load and others, are available, as for the classical OPNET scenarios in the OPNET graphical user interface, at the end of the co-simulation.

```
ESA Main loaded|--------------------------------------------------------------
|
|        Simulation and Model Library
|        Copyright 1986-2012 by
|        OPNET Technologies, Inc.
|        as a part of OPNET Release 17.1
|
|--------------------------------------------------------------
|                Application  and  Network  Performance
|--------------------------------------------------------------
|   OPNET Technologies, Inc. / 7255 Woodmont Av. / Bethesda, MD 20814, USA
|   WEB: http://www.opnet.com / TEL: +1.240.497.3000 / FAX: +1.240.497.3001
|--------------------------------------------------------------
| Protected by U.S. Patent 6,820,042.
|--------------------------------------------------------------
| Network Simulation of: ertms_tdma_proofofconcept-scenario1
|--------------------------------------------------------------

ESA Init done
ESA Load performed
Group Interfaces got
Interface at indice 0 is top.Campus Network.Mobile_1_1.cosim_intf.Trajectory
Interface at indice 1 is top.Campus Network.Mobile_1_1.cosim_intf.Message
Interface at indice 2 is top.Campus Network.Mobile_1_1.cosim_intf.Metrics
Interface at indice 3 is top.Campus Network.Mobile_1_1.cosim_intf.Train
Interface at indice 4 is top.Campus Network.Mobile_2_1.cosim_intf.Trajectory
Interface at indice 5 is top.Campus Network.Mobile_2_1.cosim_intf.Message
Interface at indice 6 is top.Campus Network.Mobile_2_1.cosim_intf.Metrics
Interface at indice 7 is top.Campus Network.Mobile_2_1.cosim_intf.Train
Interface at indice 8 is top.Campus Network.RBC1.cosim_intf.Trajectory
Interface at indice 9 is top.Campus Network.RBC1.cosim_intf.Message
Interface at indice 10 is top.Campus Network.RBC1.cosim_intf.Metrics
Interface at indice 11 is top.Campus Network.RBC1.cosim_intf.Train
Interface at indice 12 is top.Campus Network.RBC2.cosim_intf.Trajectory
Interface at indice 13 is top.Campus Network.RBC2.cosim_intf.Message
Interface at indice 14 is top.Campus Network.RBC2.cosim_intf.Metrics
Interface at indice 15 is top.Campus Network.RBC2.cosim_intf.Train
Last inactive train index 0 and RBC index 8
Opnet Manager has connected to Cosimulation Manager D-Bus interface
OPNET_MANAGER : ERTMS time is 0.004000 sec and OPNET time is 0.000000 sec and
```

Figure 5.13. *OPNET initialization traces*

```
Received Information TrainID 101 ETCSID 1 Name Train RunningNumber RUN
Connexion d'une interfaces à un ETCSID
Train with ETCSID 1 associated with interface index 0
Fin d'interconnexion de la 1eme interface sur 4 à un ETCSID
OPNET traces : Node top.Campus Network.Mobile_1_1 has been initialized as Train
Go Opnet !
OPNET traces : Node top.Campus Network.Mobile_2_1 has been initialized as Train
OPNET traces : Node top.Campus Network.RBC1 has been initialized as RBC
OPNET traces : Node top.Campus Network.RBC2 has been initialized as RBC
```

Figure 5.14. *Interfaces assignment*

```
OPNET_MANAGER : ERTMS time is 0.004000 sec and OPNET time is 0.003877 sec and difference is 0.000123
Received Trajectory ETCSID 1 XPOS 38.000000 YPOS 0.000000 Acceleration 8.000000 Speed 157.000000
OPNET traces : Node top.Campus Network.Mobile_1_1 has obtained ETCSID 1
OPNET Traces top.Campus Network.Mobile_1_1 in Cosim_Intf : obtains ETCSID 1 and has MAC address 101 (key 1) Hash has size 1
Go Opnet !
OPNET_MANAGER : ERTMS time is 0.004000 sec and OPNET time is 0.003877 sec and difference is 0.000123
Received Information TrainID 102 ETCSID 2 Name Train RunningNumber RUN
Connexion d'une interfaces à un ETCSID
Train with ETCSID 2 associated with interface index 4
Fin d'interconnexion de la 2eme interface sur 4 à un ETCSID
OPNET traces : Node top.Campus Network.Mobile_2_1 has obtained ETCSID 2
OPNET Traces top.Campus Network.Mobile_2_1 in Cosim_Intf : obtains ETCSID 2 and has MAC address 102 (key 2) Hash has size 1
Go Opnet !
OPNET_MANAGER : ERTMS time is 0.004000 sec and OPNET time is 0.003877 sec and difference is 0.000123
Received Trajectory ETCSID 2 XPOS 36.000000 YPOS 0.000000 Acceleration 19.000000 Speed 127.000000
OPNET Traces : default position of Node top.Campus Network.Mobile_2_1  of ETCSID 2: X 32260.000000 , Y 130.000000
OPNET Traces : Initial position of Node top.Campus Network.Mobile_2_1  of ETCSID 2: X 36.000000 , Y 0.000000
Go Opnet !
OPNET_MANAGER : ERTMS time is 0.004000 sec and OPNET time is 0.003877 sec and difference is 0.000123
Received Information TrainID 0 ETCSID 1234 Name RBC1 RunningNumber 12345678
Connexion d'une interfaces à un ETCSID
RBC with ETCSID 1234 associated with interface index 8
Fin d'interconnexion de la 3eme interface sur 4 à un ETCSID
OPNET traces : Node top.Campus Network.RBC1 has obtained ETCSID 1234
OPNET Traces top.Campus Network.RBC1 in Cosim_Intf : obtains ETCSID 1234 and has MAC address 1234 (key 1,234) Hash has size 1
Go Opnet !
OPNET_MANAGER : ERTMS time is 0.004000 sec and OPNET time is 0.003877 sec and difference is 0.000123
Received Trajectory ETCSID 1234 XPOS 44.000000 YPOS 0.000000 Acceleration 11.000000 Speed 58.000000
Go Opnet !
OPNET_MANAGER : ERTMS time is 0.004000 sec and OPNET time is 0.003877 sec and difference is 0.000123
Received Information TrainID 0 ETCSID 1235 Name RBC2 RunningNumber 12345679
Connexion d'une interfaces à un ETCSID
RBC with ETCSID 1235 associated with interface index 12
Fin d'interconnexion de la 4eme interface sur 4 à un ETCSID
OPNET traces : Node top.Campus Network.RBC2 has obtained ETCSID 1235
OPNET Traces top.Campus Network.RBC2 in Cosim_Intf : obtains ETCSID 1235 and has MAC address 1235 (key 1,235) Hash has size 1
Go Opnet !
OPNET_MANAGER : ERTMS time is 0.004000 sec and OPNET time is 0.003877 sec and difference is 0.000123
Received Trajectory ETCSID 1235 XPOS 44.000000 YPOS 0.000000 Acceleration 3.000000 Speed 11.000000
Go Opnet !
OPNET_MANAGER : ERTMS time is 0.500000 sec and OPNET time is 0.003877 sec and difference is 0.496123
Received Trajectory ETCSID 1 XPOS 70.000000 YPOS 0.000000 Acceleration 15.000000 Speed 11.000000
OPNET Traces : default position of Node top.Campus Network.Mobile_1_1  of ETCSID 1: X -20.000000 , Y 130.000000
OPNET Traces : Initial position of Node top.Campus Network.Mobile_1_1  of ETCSID 1: X 70.000000 , Y 0.000000
```

Figure 5.15. *Train information and initial positions setting*

```
OPNET_MANAGER : ERTMS time is 2.001000 sec and OPNET time is 2.000101 sec and difference is 0.000899
Received Trajectory ETCSID 2 XPOS 178.000000 YPOS 0.000000 Acceleration 2.000000 Speed 238.000000
OPNET Traces Current position of Node Train : X 178.000000 , Y 0.000000
Go Opnet !
OPNET_MANAGER : ERTMS time is 2.501000 sec and OPNET time is 2.000101 sec and difference is 0.500899
Received Trajectory ETCSID 1 XPOS 216.000000 YPOS 0.000000 Acceleration 4.000000 Speed 118.000000
OPNET Traces Current position of Node Train : X 216.000000 , Y 0.000000
Go Opnet !
OPNET_MANAGER : ERTMS time is 2.501000 sec and OPNET time is 2.500100 sec and difference is 0.000900
Received Trajectory ETCSID 2 XPOS 212.000000 YPOS 0.000000 Acceleration 13.000000 Speed 107.000000
OPNET Traces Current position of Node Train : X 212.000000 , Y 0.000000
Go Opnet !
OPNET_MANAGER : ERTMS time is 3.001000 sec and OPNET time is 2.500100 sec and difference is 0.500900
Received Trajectory ETCSID 1 XPOS 253.000000 YPOS 0.000000 Acceleration 15.000000 Speed 153.000000
OPNET Traces Current position of Node Train : X 253.000000 , Y 0.000000
```

Figure 5.16. *Position updates*

```
OPNET_MANAGER : ERTMS time is 10.866000 sec and OPNET time is 10.501404 sec and difference is 0.364596
Received Message MessageID 1 TimeStamp 90.700000 Train2Track ETCSIDSrc 1 ETCSIDDst 1234 CapiPrimitive 714 MessageSize 84
Data size is 80
OPNET Traces top.Campus Network.Mobile_1_1 the element at index 0 of 1 in the Hash is 101
OPNET Traces dest_nil Cosim_intf function block top.Campus Network.Mobile_1_1 (src_addr 101 key 1,234): destination is -1
Go Opnet !
OPNET Traces Mobile_1_1 : the source address in the packet is 101
OPNET Traces Mobile_1_1 : we are in From_ESYS  state in IF dest is -1 and protocol is 4321 scheduled
OPNET Traces Mobile_1_1 : we are in XMIT state sending dest is -1 controller is 0 and protocol is 4321
OPNET Traces Mobile_1_1 : we are in XMIT state sending dest is -1 controller is 0 and protocol is 4321
OPNET Traces Mobile_1_1 : we are in XMIT state sending dest is -1 controller is 0 and protocol is 4321
OPNET Traces Controller_1 : we are in From_NTWK dest -1 is replaced by 1234 and protocol 4321
OPNET Traces Controller_1 : we are in from_NTWK state of 101 sending dest is 1234 and protocol is 4321
OPNET Traces : MAC packet constructed has src 101  dest 1234 and protocol 4321
OPNET Traces top.Campus Network.node_0 : packet going to higher layer from MAC dest 1234 protocol 4321
OPNET Traces : MAC packet constructed has src 101  dest 1234 and protocol 4321
OPNET Traces : MAC packet constructed has src 101  dest 1234 and protocol 4321
OPNET Traces : MAC packet constructed has src 101  dest 1234 and protocol 4321
OPNET Traces : MAC packet constructed has src 101  dest 1234 and protocol 4321
OPNET Traces : MAC packet constructed has src 101  dest 1234 and protocol 4321
OPNET Traces : MAC packet constructed has src 101  dest 1234 and protocol 4321
OPNET Traces : MAC packet constructed has src 101  dest 1234 and protocol 4321
OPNET Traces : MAC packet constructed has src 101  dest 1234 and protocol 4321
OPNET Traces top.Campus Network.RBC1 : packet going to higher layer from MAC dest 1234 protocol 4321
OPNET Traces ARP: Received an etcs_message from MAC and sent it to ESYS
OPNET Traces top.Campus Network.Controller_4 : packet going to higher layer from MAC dest 1234 protocol 4321
OPNET Traces Controller_4 : controller received a packet from the brigde dest 1234 protocol 4321
OPNET Traces Controller_4 : we are in From_ESYS LLC proc dest is 1234 and source 0 and protocol is 4321 scheduled
OPNET Traces top.Campus Network.Controller_5 : packet going to higher layer from MAC dest 1234 protocol 4321
OPNET Traces Controller_5 : controller received a packet from the brigde dest 1234 protocol 4321
OPNET Traces Controller_5 : we are in From_ESYS LLC proc dest is 1234 and source 0 and protocol is 4321 scheduled
OPNET Traces top.Campus Network.Controller_3 : packet going to higher layer from MAC dest 1234 protocol 4321
OPNET Traces Controller_3 : controller received a packet from the brigde dest 1234 protocol 4321
OPNET Traces Controller_3 : we are in From_ESYS LLC proc dest is 1234 and source 0 and protocol is 4321 scheduled
OPNET Traces top.Campus Network.Controller_6 : packet going to higher layer from MAC dest 1234 protocol 4321
OPNET Traces Controller_6 : controller received a packet from the brigde dest 1234 protocol 4321
OPNET Traces Controller_6 : we are in From_ESYS LLC proc dest is 1234 and source 0 and protocol is 4321 scheduled
OPNET Traces top.Campus Network.Controller_2 : packet going to higher layer from MAC dest 1234 protocol 4321
OPNET Traces Controller_2 : controller received a packet from the brigde dest 1234 protocol 4321
OPNET Traces Controller_2 : we are in From_ESYS LLC proc dest is 1234 and source 0 and protocol is 4321 scheduled
OPNET Traces top.Campus Network.Controller_7 : packet going to higher layer from MAC dest 1234 protocol 4321
OPNET Traces Controller_7 : controller received a packet from the brigde dest 1234 protocol 4321
OPNET Traces Controller_7 : we are in From_ESYS LLC proc dest is 1234 and source 0 and protocol is 4321 scheduled
OPNET Traces top.Campus Network.Controller_8 : packet going to higher layer from MAC dest 1234 protocol 4321
OPNET Traces Controller_8 : controller received a packet from the brigde dest 1234 protocol 4321
OPNET Traces Controller_8 : we are in From_ESYS LLC proc dest is 1234 and source 0 and protocol is 4321 scheduled
```

Figure 5.17. *Message transfer from ERTMS simulator to OPNET for telecommunication transmission simulation*

```
OPNET Traces top.Campus Network.RBC1 : packet going to higher layer from MAC dest 1234 protocol 4321
OPNET Traces ARP: Received an etcs_message from MAC and sent it to ESYS
OPNET Traces top.Campus Network.RBC1 Cosim_intf received a packet of format etcs_message and protocol 4321
OPNET Traces top.Campus Network.RBC1 Cosim_intf (MAC 1234) : received an ETCS message from 101
OPNET Traces top.Campus Network.RBC1 : the sender_addr inside the message 1 is 101
OPNET Traces top.Campus Network.RBC1 : updated correspondance ETCSID 1  MAC 101 (key 1) and Hash size is 2
OPNET Traces top.Campus Network.RBC1 the element at index 0 of the Hash is 101
OPNET Traces top.Campus Network.RBC1 the element at index 1 of the Hash is 1234
OPNET Traces : Node top.Campus Network.RBC1 is about to update ESYS interfaces for message 1 and metrics
Opnet Manager Main received Callback for Message 1
Callback function sent a Message of ID 1 from Opnet to Manager
```

Figure 5.18. *Message whose transmission was simulated by OPNET receives an acceptance notification for routing in ERTMS simulator*

```
OPNET Traces : Node top.Campus Network.RBC2 is about to update ESYS
Opnet Manager Main received Callback for Message 90
Callback function sent a Message of ID 90 from Opnet to Manager
OPNET Traces Node top.Campus Network.RBC2 statistics are
 Total Message Received from ETCS  0.000000
Total Messages received from OPNET 15.000000
 Average delay in OPNET 0.721108 sec
```

Figure 5.19. *Metrics are updated along with message updates for each node*

5.4.2. *Efficiency of the co-simulation approach in the evaluation of railway systems*

The co-simulation approach developed in the VEGAS project and presented in this chapter achieves its main goal by introducing an accurate simulation of the telecommunication subsystem in the loop of any evaluation of the ETCS applications using an ERTMS simulator. Moreover, by concentrating all the co-simulation operations in a single process that can be included in any node model in OPNET, the co-simulation approach developed in this work can be quickly adapted in order to evaluate any other wireless technology than the GSM-R without any change in its other components (the ERTMS simulator interface and the co-simulation manager). For example, the co-simulation ESYS process can simply be added to LTE mobile equipment to obtain a train model equipped with LTE technology instead of GSM-R. The only other node that would need to be upgraded in the overall LTE infrastructure would be the one playing the role of the RBC.

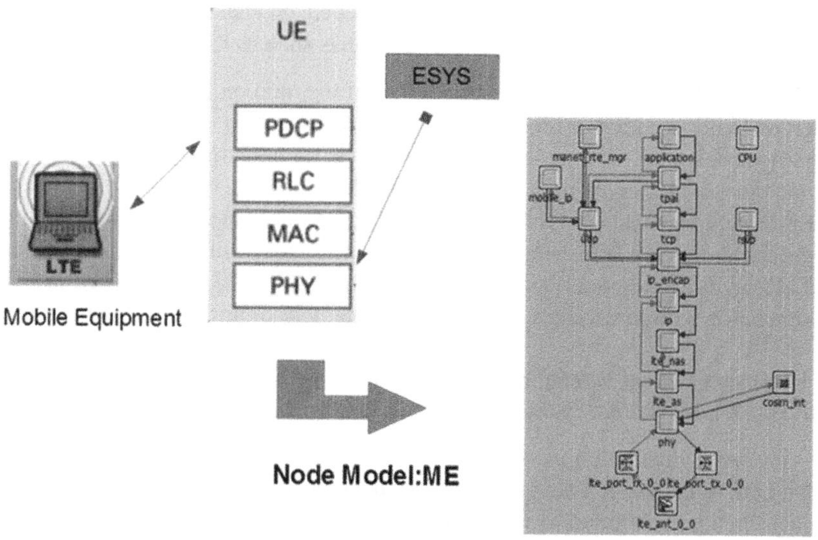

Figure 5.20. *Getting LTE (Long-Term Evolution) node models ready for co-simulation. For a color version of this figure, see www.iste.co.uk/hilt/transportation.zip*

Thus, the main advantages of the co-simulation approach developed in this work regarding railway systems evaluation, especially the ERTMS level 2, are:

– the possibility of reusing existing railway system simulators. The railway domain is subject to many rules and constraints of different types, and it involves very complicated processes that make any software development dedicated to its evaluation or exploitation a big challenge. In this context, reusing the software tools that have been already developed, tested and eventually certified is always a gain. The co-simulation approach presented in this work uses an existing professional ERTMS simulator without introducing any modification in its core functioning, by increasing its ability to interact with external tools;

– the possibility of using an improved network simulator in order to efficiently model the current telecommunication subsystem of the ERTMS and other prospective technologies that could be used in the future. As explained earlier, any operation involving the GSM-R is sent to OPNET by the co-simulation manager, which also takes into account the environment of the tracks, the network infrastructure and the movement of the trains in order to perform an accurate simulation of that operation. In this way, the feedback sent to the ERTMS simulator reflects the impact of the wireless communication in the context of the simulated ERTMS scenario. Moreover, the GSM-R infrastructure can be easily replaced by any other technology, where evaluation is needed on the same scenario, notably when evaluating potential prospective technologies such as LTE;

– the possibility of using real-world traces of the train on ERTMS level 2 lines in order to build accurate evaluation scenarios for studying the telecommunication subsystem behavior. The co-simulation manager is able to reproduce any set of chronologic events containing the movement and message information concerning an ERTMS scenario at the OPNET interface, and it operates a co-simulation without feedback to the ERTMS simulator. In this way, real-world traces of trains on ERTMS level 2 lines, such as those presented in [SON 13], can be used in offline co-simulation in order to study the behavior of the telecommunication subsystem.

However, the co-simulation approach may present some drawbacks in the development of accurate evaluation tools for railway systems, such as:

– the weak interaction possibilities due to custom simulation tools. Most of the simulation tools for railway systems are developed in the context of industrial projects that are submitted to various constraints that make them very specific. As a result, they may present some particularities such as: they are able to simulate only some modules and not the complete system; they cannot interact with other tools; they are under industrial protection and cannot be shared; and they are not event-based and cannot support pause/replay mechanisms. For example, the ERTMS simulator used in this work is not event-based, does not support pause/replay mechanisms and could not receive any feedback from external tools. It took 3 years of development to increase its ability to share some scenario data and to receive message acceptance or rejection data. However, it still does not support pause/replay and cannot operate in a tier-controlled event-based loop. As a result, an online co-simulation always depends on the ERTMS simulator scenario and can be used to evaluate the impact of this scenario on the telecommunication subsystem modeled in OPNET. The inverse situation is not possible, except through feedback on message acceptance or rejection;

– the impossibility of guaranteeing the convergence of a complete simulation scenario in online co-simulation mode. Indeed, the ERTMS simulator starts and runs without pausing until the end of the scenario. As a result, the co-simulation manager and OPNET need to operate very fast in order to send feedback before the message becomes obsolete or the default policy fixed in the configuration be applied. The sole solution to this problem is to monitor the time in both simulators and check regularly if the gap is still acceptable regarding the scenario. Once this condition is no longer satisfied, the co-simulation manager sets the default policy of the message to "accept all", stops the feedbacks to ERTMS simulator and starts an offline co-simulation with OPNET based on the events coming from the ERTMS simulator that it will continue to back up anyhow. This time synchronization problem could be avoided if the ERTMS simulator was event-based as well.

This latter observation suggests that it could be more efficient to couple in virtual laboratory, such as the one built in VEGAS project, not the simulators, but actually the models of the different components of a railway system. Indeed, multi-modeling is a well-known approach for modeling and analysis of complex systems such as railways. Moreover, many tools based on DEVS (Discrete Event System Specification) formalism [ZEI 00] have been developed in order to couple discrete and continuous models in the same virtual laboratory and generate custom simulators reproducing the entire system. The Virtual Laboratory Environment (VLE) [QUE 09], OPNET modeler and VSimRTI, to mention few, are all based on DEVS. In order to evolve from co-simulation to multi-modeling in building evaluation tools for railway systems, at least the following two major opposite facts need to be considered:

– before any custom simulation tool is designed for a component of a railway system, a model is realized first. This implies that the models of the components of railway systems are at least as available as the related simulators. Thus, building virtual laboratories on the basis of multi-modeling should be possible every time co-simulation is possible, with the advantage of avoiding the synchronization problems introduced by co-simulation;

– however, although some simulators can be made available as binary or emulated inside material devices in order to limit reverse engineering attempts, the models are more prone to intellectual property violation. In the railway domain, where these questions including confidentiality are central, it is obvious that the availability of the models that could contribute to a virtual laboratory based on multi-modeling is not guaranteed. As a result, it should be considered that developing and proposing appropriate solutions that could guarantee coupling of the models in a secure environment that preserves confidentiality and prevents intellectual property violations will be a key point in the development of future evaluation tools for the components of railway systems. Moreover, it will improve the collaboration between concurrent groups that may operate together in the design and implementation of the infrastructures for future railway systems.

5.5. Conclusion

This chapter presented a co-simulation approach developed in order to improve the joint evaluation of both the functional and telecommunication subsystems of the ERTMS by taking into account the impact of their respective behavior on each other. This approach relies on a co-simulation manager, which collects simulation data about the tracks and the movement of the trains from an ERTMS simulator implementing the functional subsystem of the ERTMS and uses them in order to simulate the transmission of the messages in OPNET. In this way, the resulting virtual laboratory has improved the evaluation of the GSM-R, and possibly other wireless technologies, by introducing realistic scenarios of the functioning of the railway system in its simulation in OPNET. Furthermore, it has improved the ERTMS simulators by introducing more realistic behavior of the telecommunication subsystem through feedback sent for any message by the co-simulation manager after it has simulated the transmission of the message through OPNET.

Despite these contributions that improve these evaluation tools, the co-simulation approach may lead to some problems, the two main ones being the limitation of interaction possibilities with custom simulators not designed for the ERTMS simulator and the impossibility of guaranteeing a co-simulation convergence due to time synchronization problems between a non-event-based simulator and event-based simulators, such as OPNET.

Prospective works are in progress to evolve from co-simulation to multi-modeling by directly coupling the models of the components instead of the resulting related simulators in order to avoid the synchronization problems. However, this approach induces new challenges concerning confidentiality which is crucial in the railway domain.

5.6. Bibliography

[AGU 07] AGUADO M., ONANDI O., JACOB E. *et al.*, *Wimax Role on CBTC Systems*, ASME/IEEE JRCICE 2007, Pueblo, 2007.

[BEU 12] BEUGIN J., MARAIS J., "Simulation-based evaluation of dependability and safety properties of satellite technologies for railway localization", *Transportation Research Part C: Emerging Technologies*, vol. 22, pp. 42–57, 2012.

[LEV 08] LEVÊQUE O., DE CICCO P., *ETCS Implementation Handbook*, Infrastructure Department, UIC, 2008.

[LOP 14] LOPEZ I., AGUADO M., JACOB E., "End-to-end multipath technology: enhancing availability and reliability in next-generation packet-switched train signaling systems", *IEEE Vehicular Technology Magazine*, vol. 9, no. 1, pp. 28–35, 2014.

[MER 07] MERA J.M., GOMEZ-REY I., CAMPOS A., "ERTMS/ETCS test simulation bench", *Urban Transport XIII Urban Transport and the Environment in the 21st Century*, UK, 2007.

[MID 08] MIDYA S., THOTTAPPILLIL R., "An overview of electromagnetic compatibility challenges in European Rail Traffic Management System", *Transportation Research Part C: Emerging Technologies*, vol. 16, no. 5, pp. 515–534, 2008.

[QUE 09] QUESNEL G., DUBOZ R., RAMAT E., "The virtual laboratory environment – an operational framework for multi-modelling, simulation and analysis of complex systems", *Simulation Modelling Practice and Theory*, vol. 17, pp. 641–653, April 2009.

[RUE 08] RUESCHE S.F., STEUER J., JOBMANN K., "The European switch – a packet-switched approach to a train control system", *IEEE Vehicular Technology Magazine*, vol. 3, no. 3, pp. 37–46, September 2008.

[SNI 14] SNIADY A., SOLER J., "LTE for railways: impact on performance of ETCS railway signaling", *IEEE Vehicular Technology Magazine*, vol. 9, no. 2, pp. 69–77, 2014.

[SON 12] SONDI P., KASSAB M., BERBINEAU M. *et al.*, "Toward a common platform for simulation-based evaluation of both functional and telecommunication subsystems of the ERTMS", *American Society of Mechanical Engineers Joint Roll Conference*, Philadelphia, pp. 351–359, 2012.

[SON 13] SONDI P., BERBINEAU M., KASSAB M. *et al.*, "Generating test scenarios based on real-world traces for ERTMS telecommunication subsystem evaluation", *International Workshop on Communication Technologies for Vehicles*, Springer-Verlag, pp. 223–231, 2013.

[UNI 10] UNISIG, System Requirements Specification, ERTMS Specifications, Subset 026 v2.3.0, ERTMS, 2010.

[ZEI 00] ZEIGLER B.P., KIM D., PRAEHOFER H., *Theory of Modeling and Simulation: Integrating Discrete Event and Continuous Complex Dynamic Systems*, Academic Press, 2000.

6

Emulating a Realistic VANET Channel in Ns-3

6.1. Introduction

Vehicular ad hoc networks are a class of MANET which have been designed to allow vehicles to exchange different types of information ranging from security messages to entertainment content. Wireless connection between nodes implies that the communication is subject to link volatility. For mainly money/time saving and efficiency reasons, the simulation of VANET is traditionally done by means of a network simulator such as Opnet, Veins, ns-2 [NS 02] or ns-3 [NS 03]. However, the plain vanilla versions of these software packages do not accurately model the main physical effects of the wireless channel [AND 06]. The consequence of this is that VANET simulations are over-optimistic. In this chapter, we will first describe the main aspects of the wireless VANET channel and underline its central role in VANET communications. Next, we will present the different modeling approaches we have followed to accurately simulate the VANET channel in ns-2 and ns-3. After showing the effect of realistic channel models on VANET simulation, we provide a solution, which is a trade-off between computing time and realism.

6.2. Influence of the channel propagation model on VANET simulation

In that which follows, we show the effect of realistic channel propagation models on Vehicular Ad-hoc Networks (VANETs) simulation. As we will see, independently from the simulator used, the more simplistic the channel propagation model used by the simulator, the worse the accuracy of the simulation. In order to

---

Chapter written by Hervé BOEGLEN, Benoit HILT and Frédéric DROUHIN.

accurately simulate the effects of the wireless channel propagation on data transmission, two elements are required. First, a physical layer, which is compliant with the transmission standard considered (i.e. IEEE802.11p or WAVE for Wireless Access in Vehicular Environment [IEE 13]) and on which it is possible to accurately apply wireless channel effects. Second, a realistic channel part obtained either from real-world measurements (statistical channel model) or by using deterministic ray-tracing software. Of course, as the mobility and the positions of the nodes have a significant impact on experienced channel conditions, we have to use a realistic mobility model too. However, this element is out of the scope of this chapter and will not be further discussed.

6.2.1. *A realistic IEEE802.11 PHY layer*

In the case of a real IEEE802.11p system, at the physical layer, the performance of a communication is evaluated with the Bit Error Rate (BER). The most accurate way to achieve this in simulation is to implement the full IEEE802.11p physical layer down to the signal level. The IEEE802.11 standard has been well described in the literature and is quite easy to implement using software packages such as IT++ [IT 16]. The IEEE802.11p standard uses packet OFDM transmission at half the rate of the IEEE802.11a standard, and hence, it can be easily implemented using an easily available IEEE802.11a code. Table 6.1 summarizes the main parameters of the two IEEE802.11a and p physical layers.

Parameters	IEEE 802.11a	IEEE 802.11p	Changes
Bit rate (Mb/s)	6, 9, 12, 18, 24, 36, 48, 54	3, 4.5, 6, 9, 12, 18, 24, 27	Half
Modulation mode	BPSK, QPSK, 16QAM, 64QAM	BPSK, QPSK, 16QAM, 64QAM	No change
Code rate	1/2, 2/3, 3/4	1/2, 2/3, 3/4	No change
Number of subcarriers	52	52	No change
Symbol duration	4 μs	8 μs	Double
Guard time	0.8 μs	1.6 μs	Double
FFT period	3.2 μs	6.4 μs	Double
Preamble duration	16 μs	32 μs	Double
Subcarrier spacing	0.3125 MHz	0.15625 MHz	Half

Table 6.1. *IEEE802.11a and p PHY parameters*

The main difficulty when integrating such a physical layer in a network simulator is the synchronization between the network simulator, which is a discrete event simulator and the continuous time signal level physical layer implementation. Moreover, we can also take into account the synchronization algorithms, which are essential in real digital communication systems. These algorithms are in general very sensitive to channel conditions [TRO 04].

6.2.2. *Accurate VANET channel propagation modeling*

6.2.2.1. *The physics of the wireless channel*

As opposed to wired network links, wireless links depending on signal propagation over the air are highly dependent on the transmission environment and are therefore highly fluctuating, especially when the nodes are moving in dense urban environments. This volatility is due to the combination of several degrading effects, which can be simulated by either deterministic or stochastic models. In what follows, we first describe the main physical effects (see Figure 6.1) we have to take into account to simulate a typical wireless channel accurately. Then, we present the particularities of a wireless VANET channel.

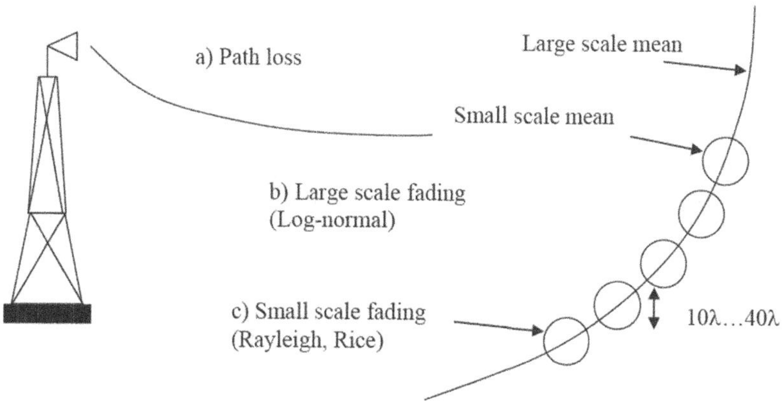

Figure 6.1. *The main wireless propagation effects*

The first encountered effect is *path loss,* which is the reduction in power density of the electromagnetic waves as they propagate through space. Path loss is described by the well-known Friis equation:

$$\alpha_p - \left(\frac{\lambda}{4\pi R}\right)$$

Attenuation α_p is therefore proportional to the square of the travelled distance R and inversely proportional to the square of the carrier wavelength $\boldsymbol{\lambda}$. This represents the first approach of a traditional link budget analysis and assumes that the transmitter and the receiver are in Line Of Sight (LOS). Hence, by knowing the transmitted power and the receiver sensitivity (i.e. the minimum power level it can handle), we can easily calculate the maximum distance of a communication at a certain frequency.

However, path loss attenuation represents the best case for a wireless communication. In practice, two important additional effects have to be taken into account, namely shadowing and small-scale fading.

The second effect appears on a large scale, typically in a few tens to a hundred wavelength units. For the 5.9 GHz frequency used by the Wireless Access in Vehicular Environments IEEE802.11p standard, this wavelength is around 5 cm. When the propagating path is obstructed by large objects, the received signal power fluctuates around its mean. This effect is called *shadowing*. Measurements have shown that power variations are best described by a lognormal distribution or alternatively by a normal distribution when expressed in logarithmic units. In practice, measurements are carried out for a specific environment, which leads to a path loss plus shadowing propagation model. For this model, the path loss attenuation is obtained as a function of the distance between the transmitter and the receiver, and a logarithmic value of the standard deviation for the shadowing effect is given.

The third effect, which is also the most important, is called *small-scale fading*. It is observed on the signal power for movements on a distance scale of a few wavelength units. This type of fading has two main physical causes. The first one comes from multi-path propagation, in which the transmitted signal interacts with the environment. Because of reflection, diffraction and diffusion interactions with the objects of the environment, the received signal is made up of several copies of the original signal called Multi-Path Components (MPC). These MPCs arrive at the receiver with different delays, amplitudes and phases and represent the so-called Channel Impulse Response (CIR). In the case of a large transmission bandwidth, the transmission signal spectrum will undergo frequency fading. In fact, this means that the spectral components (amplitude and phase) of the transmitted spectrum will be affected differently by the channel. This is called *frequency selectivity* and can be compensated for by using digital communication techniques like OFDM. The second physical effect is due to the relative movement between the transmitter and the receiver. The consequence is that the CIR becomes time-variant. This *time selectivity* can be observed on the time-domain-received signal, which is affected by amplitude fluctuations that can reach 30 dB. These amplitude fluctuations with time can be of different size depending upon the presence or absence of an LOS path in the CIR. If an LOS path is present, the fluctuations of the amplitude with time follow a Rice distribution; otherwise, they follow a Rayleigh distribution. Figure 6.2 shows an example of a received signal with and without an LOS path. We can note that the fading amplitude is greater in the absence of an LOS path. Another characteristic to be noted in Figure 6.2 is the time separation between fades: the higher the relative speed between the transmitter (TX) and the receiver (RX), the smaller the time separation. Moreover, the samples between fades are highly correlated. This is because of the Doppler effect, which produces an expansion or a compression of the transmitted spectrum. The maximum Doppler frequency shift is related to the maximum relative speed between TX and RX. From a computer implementation viewpoint, this effect is reproduced by passing the TX samples into a filter, whose bandwidth and shape are obtained from measurements. According to the type of shape (the most well-known shape is the so-called Jakes spectrum), there will be a significant impact on the reception and in particular on the BER.

Compared with mobile telephony, a VANET communication has distinct characteristics, such as fast varying surroundings, including obstructing objects, a transmitter and a receiver placed at similar heights, and a mobility that can be quite high. This leads to a great number of different situations, which imply specific channel behaviors. Since these situations are difficult to handle, the first step consists, as it was the case for mobile telephony, of identifying which may be considered as typical of VANET. Following the work on the 802.11p standard, several measurement campaigns have been conducted on typical VANET situations mainly in the United States [ACO 07, SEN 08]. The resulting channel models used to evaluate the 802.11p physical layer [ACO 07] are still the most complete implementation in terms of the number of different situations taken into account. These channels are WSSUS stochastic models and can be easily integrated in a digital communication simulator.

Other channel models proposed in the literature are more precise in the way that they take into account interacting objects in the environment by using a simplified ray-tracing approach [KAR 09]. Sometimes, this simplified ray-tracing method is tuned thanks to measurements leading to less exhaustive statistical analysis, as is the case for WSSUS stochastic models [PAS 16]. These types of channels are directly connected to the radio waves propagation physics and can, of course, model a specific VANET situation accurately. When large-scale VANET simulations have to be performed, selecting the channel model to use is an issue, as the environments encountered in a realistic simulation can be very different. In these situations, the so-called unit disk method can be used [AKH 15]. According to this method, if the receiver and the transmitter are located in a distance comprising the unit disk, communication takes place. Of course, the radius of this disk is parameterized by measurements, which have been performed for the considered situation. We refer the reader to [BOB 15] for a more detailed discussion about these VANET channel-modeling aspects.

To summarize, capturing the complexities of a vehicular channel is far from being trivial, as a large number of possible situations are encountered. Several channel models are available, which can be integrated into a network simulator, requiring more or less software development effort. However, the key element to take into account when selecting a model is to consider the trade-off between realism and simulation time. In general, the more realistic the channel model, the more time consuming the simulation. We will cover these aspects in the next sections when implementing a realistic channel model in the ns-2 and ns-3 simulators.

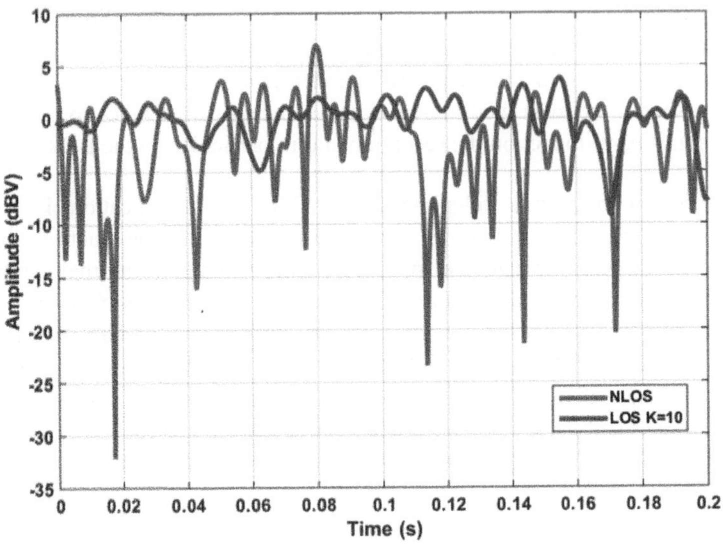

Figure 6.2. *Rayleigh (NLOS) and Rice (LOS k = 10) signal amplitude. For a color version of this figure, see www.iste.co.uk/hilt/transportation.zip*

6.3. A way to realistic channel modeling with ns-2

Ns-2 implements three different propagation models to simulate the wireless channel: the free-space model (Friis equation), the two-ray ground model and the shadowing model. As presented above, this is inadequate to provide a realistic channel model for wireless network simulation. Therefore, in order to fill this gap, the first approach we conducted was to take into account the results of deterministic ray-tracing propagation software in ns-2. The ray-tracing software computes new channel conditions for each position of the transmitter and the receiver. To maintain the simulation duration within the acceptable limits, we determine the NLOS/ LOS conditions offline using the ray-tracing tool for all possible couples of emitter/receiver into a simulation scene. For the chosen scene, that is, Munich city center, this time-consuming step took about 2 weeks for computation using more than 20 computers. Although this operation is only performed once, this is clearly not suitable for VANET simulations [HAM 09].

As indicated in section 6.2.2.1, one of the most relevant parameters when modeling a transmission channel is if the transmitter and the-receiver are in an-LOS or NLOS situation. Using this particular situation, we set up an original propagation model. It first uses a simplified ray-tracing step for the determination of the LOS/NLOS situation. As opposed to the original tool, we only take into account the distances between the transmitters and the receivers instead of their positions. This significantly reduces the pre-computation time. This information is then used as an input parameter for a statistical channel model called Spatial Channel Model Extended Urban Microcell (SCME-UM). This model has been designed by the 3GPP consortium for Urban Microcell (UM) environments [BAU 05]. The SCME model is an evolution of the SCM and extends its usage to the 2 and 5 GHz transmission bands with a transmission bandwidth of up to 100 MHz. This channel model is particularly well adapted to V2V communications, whose communication range is lower than 1,000 m. The coupling of the ray-tracing software (called CRT) [ESC 01] and the SCME-UM channel model is called UM-CRT [LED 12], which is shown in Figure 6.3. In the figure, the gray arrows show how the deterministic ray-tracing simulator (i.e. CRT) can be used with ns-2; the black arrows show how the statistical model (i.e. SCME-UM) can be used with ns-2 and the dashed lines show how to combine both into a hybrid model, which considers distances rather than positions. UM-CRT was validated by a BER comparison with the ray-tracing software (Figure 6.4). It is important to note that the use of node-to-node distances instead of positions decreases the computation time significantly, making the model more suitable for V2V simulations. However, it requires a pre-computation phase to identify the LOS/NLOS situation depending upon the distance between nodes in a given scene. This computation requires a few hours but has to be done only once per scene (e.g. a city center). With UM-CRT, the simulation of vehicular communications has improved in realism, but still requires a long, but compatible with a research work, simulation time.

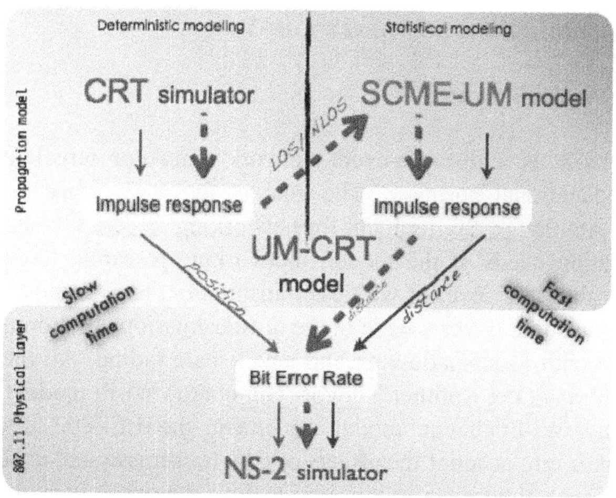

Figure 6.3. *The UM-CRT framework*

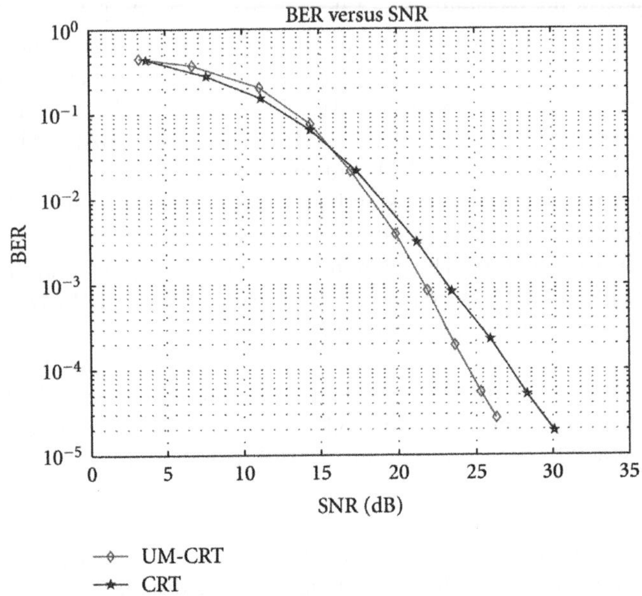

Figure 6.4. *Comparison of BER and SNR for propagation simulation that uses UM-CRT and CRT. For a color version of this figure, see www.iste.co.uk/hilt/transportation.zip*

6.4. Realistic channel modeling with ns-3

6.4.1. *The Yans WiFi model*

Ns-3, like ns-2, is a discrete-event network simulator targeted primarily for research and educational purposes. The goal of the ns-3 project is to develop a preferred open-simulation environment for networking research: it should be aligned with the simulation needs of the modern networking research. Regarding realism, and specifically the PHY level of wireless transmission, ns-3 proposes more than 25 propagation models that cover a wide range of transmission conditions. It can model phenomena like path loss, shadowing and small-scale fading. All these models are usable with the Yans (Yet Another Network Simulator) Wi-Fi model [LAC 06]. The plain vanilla Yans Wi-Fi channel model implements the IEEE802.11 standards. This means that it takes into account the effects of data transmission at different rates (i.e. digital modulations), the transmission of signaling messages in digital modulation, etc. It also makes it possible to compute the channel transmission properties, such as signal-to-noise ratio (SNR) and packet error rates (PER). As Yans Wi-Fi handles entities that are data packets, it can only provide PER information for correctly received packets. It is important to note that as the Yans Wi-Fi model is an open source, it is quite easy to add new models that are found in the literature.

Figure 6.5 shows a plot of the PER versus the SNR from a simulation with one static car and a second overtaking it. The travel speed was 15 m/s and the transmission power was 0 dBm. For these conditions, at the beginning and the end of the simulation, the cars are out of range. The propagation model was built with the following models: Friis for path loss and Rayleigh and Jakes for small-scale fading.

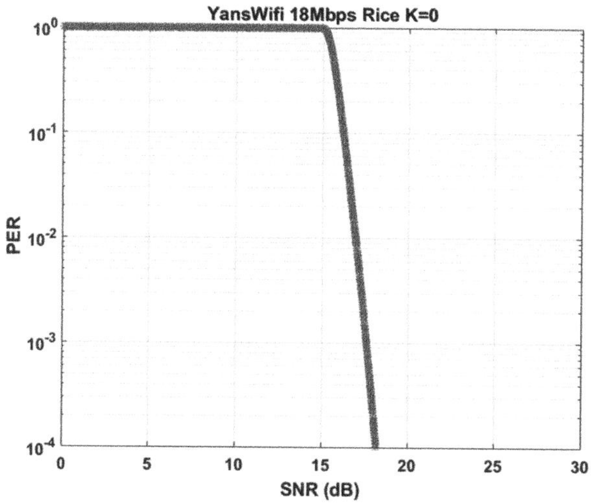

Figure 6.5. *PER versus SNR coming from Yans Wi-Fi modeled transmission*

6.4.2. *The Physim Wi-Fi model emulating OFDM-based transmission*

To go a step forward in realism, specifically with the IEEE802.11p standard dedicated to vehicular communications (Wireless Access for Vehicular Environment – WAVE), the Yans Wi-Fi model can be replaced by the Physim Wi-Fi model provided for free by the KIT as an ns-3 add-on [MIT 12]. This tool breaks the frontier between simulation and emulation as it mimics all the steps of an OFDM wireless MAC/PHY transmission. The major characteristic of Physim is that, with the help of the IT++ library [IT 13], it builds a complete digital communication physical layer going down to the OFDM packet (I and Q vectors), which is stored in a specific tag linked to the original IEEE802.11 PPDU frame. Channels effects are then applied to these OFDM packets. Because of its capability to handle bitwise information, Physim also provides SNR and BER information. Figure 6.5 shows the

BER versus SNR plot in the same conditions as those in Figure 6.6. As can be observed from the figures, there is an important difference between the two implementations. Concerning Yans Wi-Fi, there is only one PER value per SNR value. This is because the Yans Wi-Fi does not implement a full physical layer. In particular, when an SNR is calculated from a received power value, a PER value is obtained from a theoretical BER curve. On the contrary, Physim Wi-Fi implements all the signal-processing tasks related to the decoding of an OFDM packet. Within the duration of a packet, the SNR value can change and interferences from other packets can occur, thus leading to different BER values. This is the reason why different BER values are observed for the same SNR value.

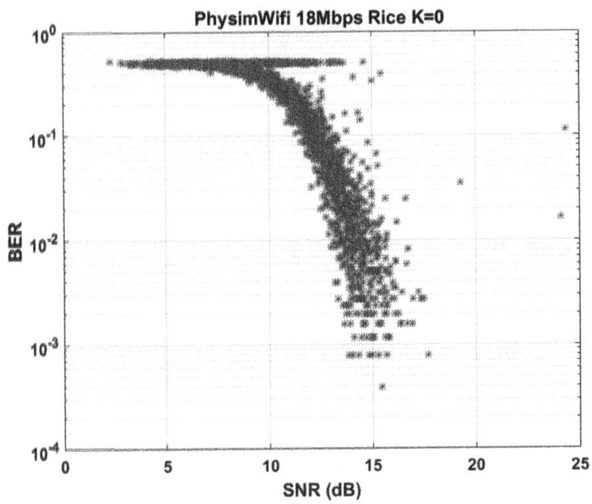

Figure 6.6. *BER versus SNR coming from Physim Wi-Fi modeled transmission. For a color version of this figure, see www.iste.co.uk/hilt/transportation.zip*

6.4.3. *Data transmission at ns-3 PHY level*

As presented in section 6.2.2, the influence of a transmission channel on transmitted data is made up of different physical phenomena. In ns-3, like for other simulation software, the channel effects on transmitted data are calculated by applying sequentially individual effects. Because of the flexibility of modern network simulators, users can customize propagation models by mixing and configuring the available propagation loss models. Channel effects are then applied by using either deterministic (e.g. Friis) or statistical (e.g. Rayleigh, Rice) methods [BEN 12]. Independently from the Wi-Fi model used (i.e. Yans or Physim), the MAC level transmits a *Packet* to the PHY level. This level, after selecting the appropriate WiFi transmission mode called *WifiMode,* which defines the characteristic elements of the transmission (coding rate, modulation, frequency, etc.), transmits the *Packet* to the "channel level" with a specific *txPower*. This channel level simulates (Yans WiFi) or emulates (Physim WiFi) all the processing steps affecting the signal used for data transmission travelling from the transmitter to the receiver. The results of the application of the channel effects to the transmitted data are then used in the reception process, which is located at the PHY level. The received data must have enough reception power to overcome the *EnergyDetectionThreshold* of the physical interface, which is typically –104 dBm for IEEE802.11p and sufficient *SNR* to be successfully decoded. This reception process also handles interference issues.

6.4.4. *The internals of WiFi channel modeling*

6.4.4.1. *The Yans WiFi case*

Yans WiFi channel models are designed to reproduce effects of over-the-air transmission disturbances on entire packets. Therefore, some effects that are of high importance in mobile communications cannot be accurately modeled. This is also why, as mentioned above, Yans WiFi is only able to produce PER information.

6.4.4.2. *The Physim WiFi case*

In the *PhySimWifiPhy::SendPacket* method, Physim WiFi packetizes the data to be sent into OFDM packets using the IT++ library. It results in an OFDM suite of samples, including a fixed preamble, training symbols, base-rate payload and full-rate payload, represented by their I and Q vectors. Interestingly, this frame could be sent out "as is" over software-defined radio (SDR) equipment. The channel effects can therefore be applied in a more effective manner than that done with Yans WiFi. With Yans WiFi, statistical models are applied packetwise through random draws in dedicated statistical distributions (e.g. Erlang/gamma random variable for the Nakagami model). In doing so, some important properties linked to packet reception, like SNR or PER, are also computed with an *InterferenceHelper* in a packetwise manner, which approximates their values. On the contrary, Physim WiFi applies the channel effects on each the OFDM sample and provides very accurate calculated SNR and BER values from different parts or all of the messages while also taking into account signal overlapping (interferences).

6.5. Case studies: emulation of realistic VANET channel models in ns-3

In this section, we will show how to take advantage of ns-3 models in order to simulate a realistic VANET channel. The first implementation can be realized with both Yans and Physim WiFi, and it is interesting to compare them. The second implementation, which implies multi-path propagation, can only be realized with Physim WiFi. This latter case is particularly interesting because, on the one hand, we will show how to use normalized statistical channel models, including those designed for VANET, and on the other hand, we will underline the limits that have to be taken into account when performing these types of simulation.

6.5.1. *A simplified VANET channel model for an urban environment*

One thing to keep in mind when using channel models is that their validity is limited to a specific environment. To be clear, one cannot use a highway channel model to draw conclusions for an urban environment. As discussed earlier, a wireless propagation channel model has to reproduce path loss, shadowing and small-scale fading efficiently. As a first yet realistic approach, let us implement a simple V2V urban channel model. To be able to compare Yans WiFi and Physim WiFi models, we will not take into account the frequency selectivity in this model, that is, the CIR will be made of only one path. Moreover, as this is a common situation in VANET, we will consider an LOS condition between the transmitter and the receiver.

In order to more precisely analyze the effect of the channel model used, we will consider two nodes. The first one (N2) remains fixed and is located 500 m away from the second one (N1). This second node transmits its data to N2 and travels toward it, passes by and moves away up to a distance of 500 m (Figure 6.7). N2 travels at a speed of 14 m/s (≈50 km/h).

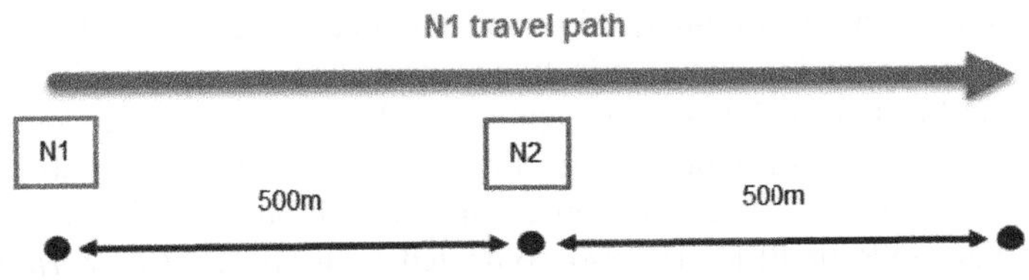

Figure 6.7. *Nodes configuration for simulation*

To model path loss and shadowing effects, we consider a model from the literature, the shadow-fading model for V2V environments of Abbas *et al.* [ABB 15] which we add to both Yans and Physim WiFi. This model considers two types of environments, namely Highway and Urban, and two main situations: LOS and OLOS. Obstructed LOS (OLOS) is a situation when the LOS path gets obstructed completely or partially by another vehicle [ABB 15]. This is different from an NLOS situation when, for example, a building blocks the LOS path completely.

Concerning small-scale fading, we consider an LOS path with K = 3 and model the Doppler spread using a Jakes filter. The simulation parameters are summarized in Table 6.2.

Path loss and shadowing	Abbas *et al.* Shadow-fading model
Small-scale fading	One Rice path + Jakes Doppler filter
Packet size	256 bytes
Interpacket interval	0.1 s
Data rate	18 Mbps (16 QAM, R = 3/4)
Transmission power	+20 dBm
Maximum distance between N1 and N2	500 m
Simulation time	90s

Table 6.2. *Simple Yans WiFi propagation channel model simulation parameters*

After averaging over 10 simulations, we obtain a PDR result for Yans WiFi that is 72 and 89% for Physim WiFi. This observed difference comes from the fact that Yans WiFi considers that if the received power threshold is exceeded, a packet is received, whereas Physim WiFi implements a full 802.11p reception. When Physim WiFi considers the reception power sufficient, it carries out a full decoding of the packet, which can fail for several reasons (e.g. SNR fluctuates on the duration of the packet). In this particular situation, the packet is dropped even if the reception power is above the prerequisite threshold.

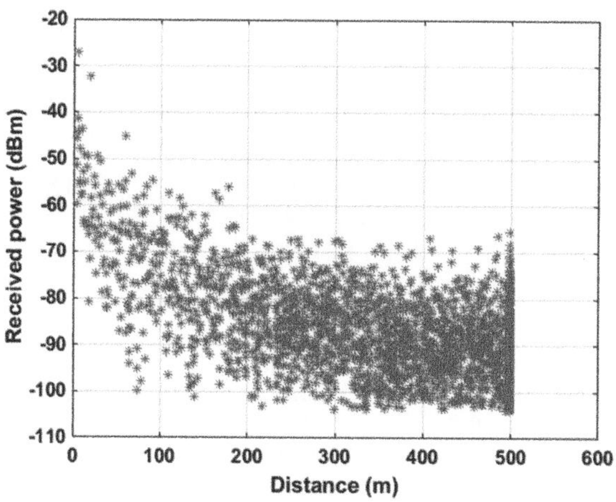

Figure 6.8. *Yans WiFi received power versus distance. For a color version of this figure, see www.iste.co.uk/hilt/transportation.zip*

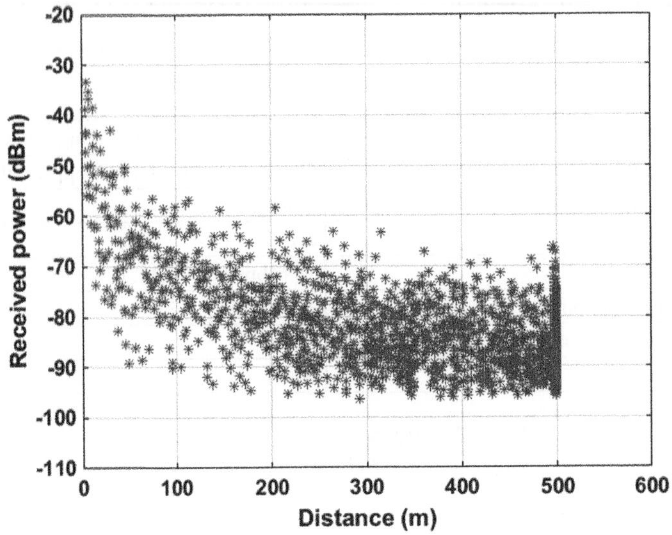

Figure 6.9. *Physim WiFi received power versus distance. For a color version of this figure, see www.iste.co.uk/hilt/transportation.zip*

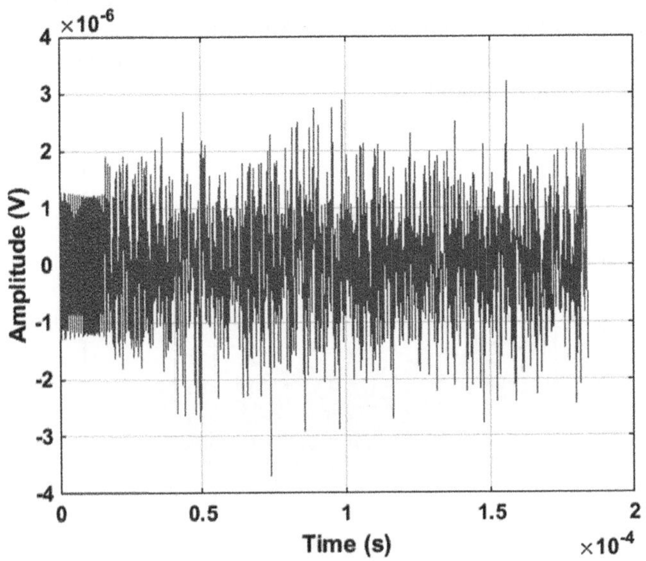

Figure 6.10. *An example of a received OFDM packet coming from Physim WiFi in ns-3. For a color version of this figure, see www.iste.co.uk/hilt/transportation.zip*

With the same channel model, we can note that the Yans WiFi is 15% more pessimistic than the Physim WiFi. If we compare the simulation time, we get 2 s for Yans WiFi and 22 s for Physim WiFi. This is expected, as Physim WiFi also implements a full IEEE802.11p physical layer up to the OFDM sample. Let us now go a step further and add a complete multi-path small-scale statistical channel model.

6.5.2. *A normalized VANET channel model for urban environments*

In what follows, we will use one of the channel models published in [ACO 07]. These channel models represent typical VANET environments and have been used as reference models to evaluate the IEEE802.11p physical layer. These models are part of the Physim WiFi implementation. It is important to note that it is not possible to use these models with the plain vanilla ns-3, as the Yans WiFi model does not model the physical layer down to the signal level. The simulation parameters are summarized in Table 6.3. As we will see in the following, these models require quite a high computation time. For this reason, we implement the ITU Vehicular A channel model (Table A.2.6.3 of [JAI 07]), which does not use the same computing method for the calculation of channel coefficients. Both channel models give similar results in terms of PDR. However, while the simulation with the Acosta-Marum channel model requires about 360 min, the ITU Vehicular A one lasts only 6 min (measured on an Intel Core I7 2600K platform)! This huge difference is mainly due to the method used to calculate the Doppler spectrum. For the ITU model, it is obtained by the so-called filtering method, whereas for the Acosta-Marum channel, it is calculated by using a time-consuming IFFT algorithm. This method is needed because VANET Doppler spectra have shapes quite different from those of the classical ones.

Path loss and shadowing	Abbas *et al.* Shadow-fading model
Small-scale fading	Acosta-Marum V2V Urban Canyon Oncoming or ITU Vehicular A
Packet size	256 bytes
Interpacket interval	0.1 s
Data rate	6 Mbps (QPSK, R = 1/2), 18 Mbps (16QAM, R = 3/4)
Transmission power	+20 dBm
Max. distance between N1 and N2	500 m
Simulation time	90 s

Table 6.3. *Simple Physim WiFi propagation channel model simulation parameters*

In addition to the PDR and the computation time, Figures 6.11 and 6.12 show the BER as a function of the SNR for the data rates selected (6 and 18 Mbps) and for the two channel models considered. There are several interesting things to note. The first one is that we recognize the expected shapes of the BER curves for the different modulations used. The Physim WiFi implementation really mimics a full 802.11p receiver. Moreover, we can also note that there are no BER values lower than 8.10^{-4}. This is simply due to the properties of the convolutional code, which correct all the errors when the BER is lower than this value. Considering the results in terms of PDR, they are very similar for the 18 Mbps case. Thus, both channel models give a PDR of about 20%. However, for the QPSK modulation (6 Mbps), the Acosta-Marum channel model gives 20% PDR, whereas the ITU Vehicular A is less selective, giving a PDR of 40%. This difference can be explained by the particular shape of the Acosta-Marum channel model Doppler spectrum.

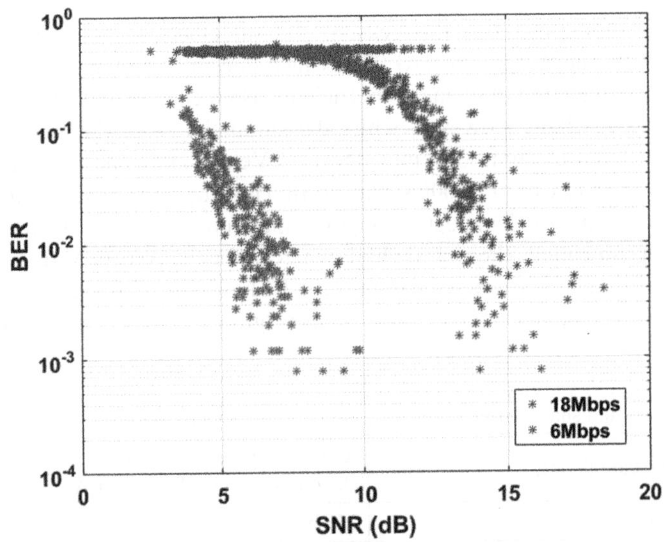

Figure 6.11. *BER versus SNR with Acosta-Marum. For a color version of this figure, see www.iste.co.uk/hilt/transportation.zip*

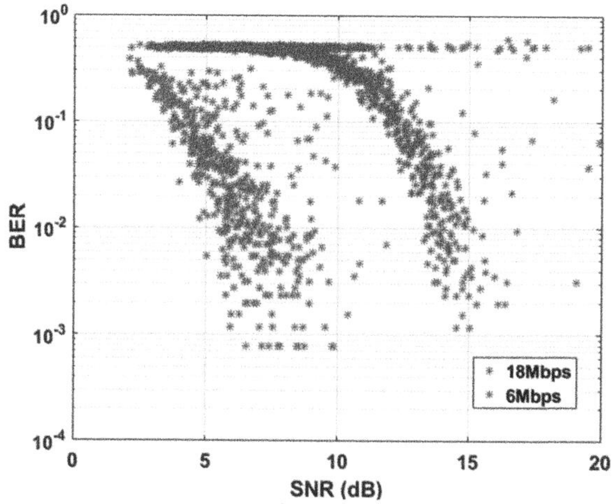

Figure 6.12. *BER versus SNR with IUT Vehicular A. For a color version of this figure, see www.iste.co.uk/hilt/transportation.zip*

6.6. Conclusion and discussion

Channel propagation modeling is a broad topic in the wireless communication domain. In this area, ray-launching or ray-tracing propagation simulators are widely used. These tools require a high degree of precision in the description of the channel environment (buildings, architectural elements, trees, advertising hoardings, etc.) in order to compute the CIR accurately. They can accurately reproduce the main propagation mechanisms but require a high computing power, thus excluding them as a possible tool in wireless networking simulations. Because of this, numerous approaches were explored to simulate channel propagation in a fast yet efficient way. If the large- and medium-scale effects (i.e. path loss and shadowing) are easy to model for vehicular environments (i.e. urban, suburban and highway) through deterministic or statistical models, things become more complex when modeling has to handle small-scale fast fading effects.

Providing statistical and realistic models of these effects requires the collection of data from intensive measurements campaigns. This is, of course, time consuming and very expensive. These campaigns were performed for the deployment of cellular telephony. Nowadays, the resulting models are well known and freely available. Concerning channel models for VANET simulations, the task is not so easy, as the variety of situations encountered is large. However, most of the physical specifications have been identified by several research works. The main question that remains is: does taking all these aspects into account in a network simulator really matter?

One way to lower the computation resources required by a full ray-launching propagation simulation is to reduce the number of effects taken into account and to compute only the key propagation elements. In addition, regarding the power of the propagated signal, one of the most important characteristics to consider is the LOS or NLOS situation. In order to determine if the ray between an emitter and a receiver is in an LOS or an NLOS situation, we have used a simplified version of the Urban Microcell (UM) statistical channel model. From these two elements, we created a semi-deterministic propagation model merging the simplified pre-computed ray-launching model and the UM LOS/NLOS situation determination. This model, called UM-CRT, offers a high degree of realism and quite low computer power requirements. This makes it suitable for realistic wireless network simulation of a few tens of nodes; however, it is not suitable for large VANET simulations of more than a hundred nodes.

In order to further reduce the simulation time[1], we have explored two 802.11 models for the ns-3 simulator. The default WiFi model is called Yans WiFi [LAC 06]. We have shown that it offers quite a good level of realism and a low computation time but it is limited when one wants to add channel models due to the lack of a realistic physical layer. Alternatively, the Physim WiFi model, which is an additional module for ns-3 provided by the KIT [MIT 12], emulates all the steps of real 802.11p receiver down to the OFDM level. This feature allows the easy integration of statistical channel models. In the VANET area, the most well-known is the set of Acosta-Marum statistical channel models [ACO 07]. These were used to evaluate the IEEE802.11p/WAVE physical layer during the IEEE standardization phase. We have used one of them as a reference in this chapter. It has been observed that it requires a very long computation time mainly due to the method used to calculate the Doppler spectrum. In order to lower the processing time, we looked for an alternative channel model. The best candidate we have found is the ITU Vehicular A statistical channel model [JAI 07]. Our evaluation shows that for the transmission rates using 16 QAM and 64 QAM digital modulations, the results in terms of PDR and BER are similar to those of the Acosta-Marum implementation. Concerning the simulation duration, the gain is significant: 6 min as opposed to 360 min for Acosta-Marum. For lower rates (i.e. BPSK and QPSK modulations), the difference between the two channel models is 20% (Acosta-Marum channel gives the worst results, i.e. a PDR of 20%). Finally, as far as wireless networking is concerned (and VANET, in particular), we have to bear in mind that the wireless channel model is essential. As opposed to wired networking, and especially when testing new routing algorithms in high-speed VANET situations, the user has to expect PDRs in the order of 40% at the best. Moreover, and especially for VANET, the communication links have stability times than can be quite low (in the order of 10 ms).

1 A couple of hours for a simulation of a hundred nodes in a few square kilometers area.

6.7. Appendix A: The Abbas et al. Model Implementation

This appendix provides the source code of the Abbas [ABB 15] path loss model for Vehicle-to-Vehicle network simulators for both the Yans and Physim WiFi models. The integration of this code in ns-3 will add a new model in the list of propagation models. Such a list looks like A) for Yans and B) for Physim WiFi models. It is important to note that the parameters of the Abbas LOS propagation model have been extracted from the text and Table 2 of [ABB 15].

A)
```
YansWifiChannelHelper wifiChannel;
wifiChannel.SetPropagationDelay("ns3::ConstantSpeedPropagationDelayMo
del");

// -- Setting a Abbas LOS propagation loss. Replaces Friis plus
// statistic shadowing in vehicular situation
wifiChannel.AddPropagationLoss ("ns3::TwoRandShadowingLineOfSight
PropagationLossModel",
 "Distance0", DoubleValue (10.0), // values based on reference paper
 "Distanceb", DoubleValue (104),
 "PathLoss0", DoubleValue (-63.9),
 "ExponentN1", DoubleValue (-1.81),
 "ExponentN2", DoubleValue (-2.85),
 "SigmaM1", DoubleValue (4.15),
 "SigmaM2", DoubleValue (4.15) );

// -- Setting Nakagami propagation loss for Rice
wifiChannel.AddPropagationLoss ("ns3::NakagamiPropagationLoss
Model",
 "Distance1", DoubleValue (0.0),
 "Distance2", DoubleValue (0.0),
 "m0", DoubleValue (0.0),
 "m1", DoubleValue (0.0),
 "m2", DoubleValue (5.76) );
```

B)
```
PhySimWifiChannelHelper wifiChannel;
wifiChannel.SetPropagationDelay ("ns3::ConstantSpeedPropagation
DelayModel");

// -- Setting dummy propagation loss - Mandatory in PhySim
wifiChannel.AddPropagationLoss ("ns3::PhySimPropagationLossModel" );

// -- Setting Abbas loss propagation loss. Replaces Friis plus
// statistic shadowing in vehicular situation
wifiChannel.AddPropagationLoss ("ns3::PhySimTwoRandShadowingLine
OfSightPropagationLoss",
 "Distance0", DoubleValue (10.0), // values based on reference paper
 "Distanceb", DoubleValue (104.0),
 "PathLoss0", DoubleValue (-63.9),
 "Exponent1", DoubleValue (-1.81),
 "Exponent2", DoubleValue (-2.85),
```

```
 "Sigma1", DoubleValue (4.15),
 "Sigma2", DoubleValue (4.15) );

// Calculation of the norm Doppler
double nodeSpeed=10.0;
double lineOfSightDoppler = (nodeSpeed*5.9e9)/(0.3e9*10e6);

// -- Setting a Rice propagation loss
wifiChannel.AddPropagationLoss ("ns3::PhySimRicianPropagationLoss",
 "MinimumRelativeSpeed", DoubleValue (2.0),
 "LineOfSightPower", DoubleValue(7.0),
 "UseShortcut", BooleanValue (false),
 "LineOfSightDoppler", DoubleValue (lineOfSightDoppler) );

// -- Setting an vehicular channel propagation loss
wifiChannel.AddPropagationLoss("ns3::PhySimVehicularChannel
PropagationLoss",
"ChannelProfile", EnumValue (V2V_URBAN_CANYON_ONCOMING),
"MinimumRelativeSpeed", DoubleValue (2.0) );
```

This is the source code of the Abbas LOS propagation model for Yans C) and Physim D) WiFi models. It is important to note that Physim WiFi is currently available only for ns-3 version 3.15 and below.

```
C) NS_OBJECT_ENSURE_REGISTERED (TwoRandShadowingLineOfSight
PropagationLossModel);
TypeId TwoRandShadowingLineOfSightPropagationLossModel::GetTypeId
(void)
{
 static TypeId tid=TypeId("ns3::TwoRandShadowingLineOfSight
 PropagationLossModel")
 .SetParent<PropagationLossModel> ()
 .AddConstructor<TwoRandShadowingLineOfSightPropagationLoss
   Model> ()
 .AddAttribute ("Distance0",
   "1st breakpoint distance d0.",
   DoubleValue (10.0),
   MakeDoubleAccessor (&TwoRandShadowingLineOfSightPropagation
   LossModel::m_distance0),
   MakeDoubleChecker<double> ())
 .AddAttribute ("Distanceb", "breakpoint distance",
   DoubleValue (104), // value based on reference paper
   MakeDoubleAccessor (&TwoRandShadowingLineOfSightPropagationLoss
   Model::m_distanceb),
   MakeDoubleChecker<double> ())
 .AddAttribute ("PathLoss0", "Free space path loss plus the
 accumulative antenna gain (PLf + Ga)",
   DoubleValue (-56.5),
   MakeDoubleAccessor(&TwoRandShadowingLineOfSightPropagation
   LossModel::m_pathLoss0),
   MakeDoubleChecker<double> ())
 .AddAttribute ("ExponentN1", "Path loss exponent until the
   breakpoint distance b",
   DoubleValue (-1.81),
   MakeDoubleAccessor(&TwoRandShadowingLineOfSightPropagation
   LossModel::m_exponentN1),
   MakeDoubleChecker<double> ())
```

```
    .AddAttribute ("ExponentN2", "Path loss exponent after the
      breakpoint distance b",
      DoubleValue (-2.85),
      MakeDoubleAccessor (&TwoRandShadowingLineOfSightPropagation
      LossModel::m_exponentN2),
      MakeDoubleChecker<double> ())
    .AddAttribute ("SigmaM1", "Standard deviation until the
      breakpoint distance b",
      DoubleValue (4.15),
      MakeDoubleAccessor (&TwoRandShadowingLineOfSightPropagation
      LossModel::m_sigmaM1),
      MakeDoubleChecker<double> ())
    .AddAttribute ("SigmaM2", "Standard deviation after the
      breakpoint distance b",
      DoubleValue (4.15),
      MakeDoubleAccessor (&TwoRandShadowingLineOfSightPropagation
      LossModel::m_sigmaM2),
      MakeDoubleChecker<double> ())
    .AddAttribute ("NormalRvM1", "Access to the underlying
      NormalRandomVariable for M1",
      StringValue ("ns3::NormalRandomVariable"), MakePointerAccessor
     (&TwoRandShadowingLineOfSightPropagation
      LossModel::m_normalRvM1),
      MakePointerChecker<NormalRandomVariable> ())
    .AddAttribute ("NormalRvM2", "Access to the underlying
      NormalRandomVariable for M2",
      StringValue ("ns3::NormalRandomVariable"), MakePointerAccessor
      (&TwoRandShadowingLineOfSightPropagation
      LossModel::m_normalRvM2),
      MakePointerChecker<NormalRandomVariable> ())
  ;
  return tid;
}

TwoRandShadowingLineOfSightPropagationLossModel::TwoRandShadowingLine
OfSightPropagationLossModel () {}

double TwoRandShadowingLineOfSightPropagationLossModel::DoCalcRx
Power (double txPowerDbm, Ptr<MobilityModel> a, Ptr<MobilityModel> b)
const
{
 double distance = a->GetDistanceFrom (b);
 double pathLoss ;

 if (distance < m_distance0) {
   // See paper p7 "there are only a few samples available for
   //d < 10 m, thus the validity range of the model is set to
   //d > 10 m and let d0 = 10 m"
   pathLoss = m_pathLoss0 + 10 * m_exponentN1 * log10
   (distance / m_distance0) + m_normalRvM1->GetValue (0,
   (m_sigmaM1 * m_sigmaM1));
 }
 else
   if (distance < m_distanceb) {
     pathLoss = m_pathLoss0 + 10 * m_exponentN1 * log10
```

```
    (distance / m_distance0) + m_normalRvM1->GetValue (0, (m_sigmaM1
* m_sigmaM1));
   }
   else {
     pathLoss = m_pathLoss0 + 10 * m_exponentN1 * log10
     (m_distanceb / m_distance0) + 10 * m_exponentN2 * log10
     (distance / m_distanceb) + m_normalRvM2->GetValue
     (0, (m_sigmaM2 * m_sigmaM2)) ;
 }

 NS_LOG_DEBUG ("TwoRandShadowingLineOfSightPropagation
 LossModel::DoCalcRxPower() pathLoss = " << pathLoss);

 return txPowerDbm + pathLoss;
}

int64_t
TwoRandShadowingLineOfSightPropagationLossModel::DoAssignStreams
(int64_t stream)
{
 m_normalRvM1->SetStream (stream);
 m_normalRvM2->SetStream (stream);
 return 2;
}
```

D) `NS_OBJECT_ENSURE_REGISTERED (PhySimTwoRandShadowingLineOfSight`

```
PropagationLoss);
TypeId PhySimTwoRandShadowingLineOfSightPropagation
Loss::GetTypeId (void)
{
 static TypeId tid = TypeId("ns3::PhySimTwoRandShadowing
 LineOfSightPropagationLoss")
 .SetParent<PhySimPropagationLossModel> ()
 .AddConstructor<PhySimTwoRandShadowingLineOfSight
 PropagationLoss> ()
 .AddAttribute ("Distance0", "1st breakpoint distance d0.",
   DoubleValue (10.0),
   MakeDoubleAccessor (&PhySimTwoRandShadowingLineOfSight
   PropagationLoss::m_distance0),
   MakeDoubleChecker<double> ())
 .AddAttribute ("Distanceb", "breakpoint distance",
   DoubleValue (104), // value given based on reference paper
   MakeDoubleAccessor (&PhySimTwoRandShadowingLineOfSight
   PropagationLoss::m_distanceb),
   MakeDoubleChecker<double> ())
 .AddAttribute ("PathLoss0", "Free space path loss plus the
   accumulative antenna gain (PLf + Ga)",
   DoubleValue (-72.3),
   MakeDoubleAccessor (&PhySimTwoRandShadowingLineOfSight
   PropagationLoss::m_pathLoss0),
   MakeDoubleChecker<double> ())
 .AddAttribute ("Exponent1", "Path loss exponent until the
   breakpoint distance b",
   DoubleValue (-1.81),
   MakeDoubleAccessor (&PhySimTwoRandShadowingLineOfSight
   PropagationLoss::m_exponentn1),
   MakeDoubleChecker<double> ())
```

```
  .AddAttribute ("Exponent2", "Path loss exponent after the
    breakpoint distance b",
    DoubleValue (-2.85),
    MakeDoubleAccessor (&PhySimTwoRandShadowingLineOfSight
    PropagationLoss::m_exponentn2),
    MakeDoubleChecker<double> ())
  .AddAttribute ("Sigma1", "Standard deviation until the
    breakpoint distance b",
    DoubleValue (6.67),
    MakeDoubleAccessor (&PhySimTwoRandShadowingLineOfSight
    PropagationLoss::SetSigma1, &PhySimTwoRandShadowingLineOfSight
    PropagationLoss::GetSigma1),
    MakeDoubleChecker<double> ())
  .AddAttribute ("Sigma2", "Standard deviation after the
    breakpoint distance b",
    DoubleValue (6.67),
    MakeDoubleAccessor (&PhySimTwoRandShadowingLineOfSight
    PropagationLoss::SetSigma2, &PhySimTwoRandShadowingLineOfSight
    PropagationLoss::GetSigma2),
    MakeDoubleChecker<double> ())
  ;
 return tid;
}

PhySimTwoRandShadowingLineOfSightPropagationLoss::PhySimTwoRand
ShadowingLineOfSightPropagationLoss () : PhySimPropagationLossModel
(), m_normalVariableSigma1 (0), m_normalVariableSigma2 (0) {}

PhySimTwoRandShadowingLineOfSightPropagationLoss::~PhySimTwoRandShado
wingLineOfSightPropagationLoss () {}

void PhySimTwoRandShadowingLineOfSight
PropagationLoss::DoCalcRxPower (Ptr<PhySimWifiPhyTag> tag,
Ptr<MobilityModel> a, Ptr<MobilityModel> b) const
{
 double distance = a->GetDistanceFrom (b);
 double pathLoss ;
 double randValue = 0 ;

 if (distance < m_distanceb) {
   randValue = m_normalVariableSigma1->GetValue () ;
   NS_LOG_DEBUG("randValue1 = " << randValue);
   ("PhySimTwoRandShadowingLineOfSightPropagationLoss::
   DoCalcRxPower m_normalVariableSigma1 = " << randValue);
   // See paper p7 "there are only a few samples available for
   //d < 10 m, thus the validity range of the model is set to
   //d > 10 m and let d0 = 10 m"
   pathLoss = m_pathLoss0 + 10 * m_exponentn1 * log10 (distance /
   m_distance0) + randValue ;

 }
 else {
   randValue = m_normalVariableSigma2->GetValue () ;
   NS_LOG_UNCOND("randValue2 = " << randValue);
   pathLoss = m_pathLoss0 + 10 * m_exponentn1 * log10 (m_distanceb
   / m_distance0) + 10 * m_exponentn2 * log10 (distance
```

```
    / m_distanceb) + randValue ;
  }

  itpp::cvec input = tag->GetRxedSamples ();
  itpp::cvec output = input * sqrt (pow (10, pathLoss / 10.0));
  tag->SetPathLoss (pathLoss);
  tag->SetRxSamples (output);
}
```

6.8. Bibliography

[ABB 15] ABBAS T., SJÖBERG K., KAREDAL J. *et al.*, "A measurement based shadow fading model for vehicle-to-vehicle network simulations", *International Journal of Antennas and Propagation*, vol. 2015, Article ID 190607, 2015.

[ACO 07] ACOSTA-MARUM G., INGGRAM M.A., "Six time- and frequency- selective empirical channel models for vehicular wireless LANs", *IEEE Vehicular Technology Magazine*, vol. 2, no. 4, pp. 4–11, December 2007.

[AKH 15] AKHTAR N., ERGEN S.C., OZKASAP O., "Vehicle mobility and communication channel models for realistic and efficient highway VANET simulation", *IEEE Transactions on Vehicular Technology*, vol. 64, no. 1, pp. 248–262, January 2015.

[AND 06] ANDEL T.R., YASINSAC A., "On the credibility of MANET simulations", *Computer*, vol. 39, no 7, pp. 48–54, 2006.

[BAU 05] BAUM D.S., SALO J., DEL GALDO G. *et al.*, "An interim channel model for beyond-3G systems", *Proceeding of IEEE VTC'05*, Stockholm, pp. 3132–3136, May 2005.

[BEN 12] BENIN J., NOWATKOWSKI M., OWEN H., "Vehicular network simulation propagation loss model parameter standardization in ns-3 and beyond", *Southeastcon, 2012 Proceedings of IEEE*, IEEE., pp. 1–5, March 2012.

[BOB 15] BOBAN M., VIRIYASITAVAT W., "Channel Models for Vehicular Communications", in CAMPOLO C., MOLINARO A., SCOPIGNO R. (eds), *Vehicular ad hoc Networks*, Springer, 2015.

[CAM 15] CAMPOLO C., *Vehicular Ad hoc Networks*, Chapter 12, Springer, 2015.

[ESC 01] ESCARIEU F., POUSSET Y., AVENEAU L. *et al.*, "Outdoor and indoor channel characterization by a 3D simulation software" *Proceedings of the 12th International Symposium on Personal, Indoor and Mobile Radio Communications (PIMRC'01)*, Boston, pp. B105–B111, October 2001.

[HAM 09] HAMIDOUCHE W., VAUZELLE R., OLIVIER C. *et al.*, "Impact of realistic MIMO physical layer on video transmission over mobile ad hoc network", *Proceedings of the IEEE 20th Personal, Indoor and Mobile Radio Communications Symposium (PIMRC '09)*, Tokyo, pp. 187–191, September 2009.

[IEE 13] IEEE 1609, "Family of standards for Wireless Access in Vehicular Environments (WAVE)", U.S. Department of Transportation, April 13, 2013.

[IT 13] *IT++* Library of Mathematical, Signal Processing and Communication Classes and Functions, http://itpp.sourceforge.net/4.3.1/, 2013.

[JAI 07] JAIN R., "Channel Models: a Tutorial", *WiMAX forum AATG*, pp. 1–6, February 2007.

[KAR 09] KAREDAL J. TUFVESSON F., CZINK N. *et al.*, "A geometry-based stochastic MIMO model for vehicle-to-vehicle communications", *IEEE Transactions. Wireless Communications*, vol. 8, no. 7, pp. 3646–3657, July 2009.

[LAC 06] LACAGE M., HENDERSON T.R., "Yet another network simulator", *Proceeding from the 2006 workshop on ns-2: the IP network simulator*, ACM, p. 12, 2006.

[LED 12] LEDY J., BOEGLEN H., POUSSARD A.-M. *et al.*, "A semi-deterministic channel model for VANETs simulations", *Hindawi International Journal of Vehicular Technology*, vol. 2012, Article ID 492105, 2012.

[MIT 12] MITTAG J., available at: https://dsn.tm.kit.edu/ns3-physim.php, 2012.

[NS 02] The ns-2 discrete event simulator targeted at networking research, available at: http://nsnam.sourceforge.net/wiki/index.php/Main_Page, 2002.

[NS 03] The ns-3 discrete-event network simulator for Internet systems, available at: https://www.nsnam.org, 2003.

[PAS 16] PASCHALIDIS P. NUCKELT J., MAHLER K. *et al.*, "Investigation of MPC correlation and angular characteristics in the vehicular urban intersection channel using channel sounding and ray tracing", *IEEE Transactions on Vehicular Technology*, vol. 65, no. 8, pp. 5874–5886, August 2016.

[SEN 08] SEN I., MATOLAK D., "Vehicle-to-vehicle channel models for the 5-GHz band", *IEEE Transactions on Intelligent Transportation Systems*, vol. 9, no. 2, pp. 235–245, June 2008.

[TRO 04] TROYA CHINCHILLA A., "Synchronization and Channel Estimation in OFDM: Algorithms for Efficient Implementation of WLAN Systems", PhD Thesis, Brandenburgischen Technischen Universität Cottbus, 2004.

7

CONVAS: Connected Vehicle Assessment System for Realistic Co-simulation of Traffic and Communications

7.1. Introduction

Connected vehicle technology enables vehicles to communicate with each other and the infrastructure wirelessly. Automated vehicle technology senses the driving environment and operates a vehicle with limited or even without human input. These technologies together provide a platform for creating a wide array of applications to address real-world problems of how to assist and improve mobility, safety and the environment through next generation Intelligent Transportation Systems (ITS). With a limited initial market penetration, emerging technology components and unknown human behavioral responses, we view realistic simulation as a very powerful and cost-effective method for testing, developing and evaluating various components of these new technologies.

Traditional traffic simulation focuses on microscopic road behaviors, simple interactions between vehicles and interactions with the transportation infrastructure. Communications are often assumed to be ideal or are crudely simplified for modeling purposes. In many cases, communication effects are post-processed using detailed vehicle trajectories and do not affect the traffic simulation. The major limitation of such approaches is that simulated vehicles make no adjustments based on the packets received or wireless reception characteristics. Integrating both traffic and communication simulations at a fine time scale for large simulations becomes a challenge. Furthermore, realistic wireless communication models for traffic simulation become increasingly necessary for evaluating the impacts of new

Chapter written by Justinian ROSCA, Ines UGALDE, Praprut SONGCHITRUKSA and Srinivasa SUNKARI.

technologies especially for time-critical applications. For example, there is no widely agreed upon physical model for Dedicated Short Range Communications (DSRC) over the 75 MHz of spectrum (5850 to 5925 MHz) using the IEEE 802.11p standard. This spectrum was allocated by the FCC, and standards including IEEE 802.11p (Physical and MAC layers), IEEE 1609.1-4 (upper protocols) and SAEJ2735/2945.1 (application layer, targeting vehicular applications) have been developed to support ITS [KEN 11]. The number of variables to take into account is countless: communication frequency, vehicle movement, antenna mounting and type, speed of transmission, traffic density, environment type (from urban to highway), weather, to name just a few. There exist several advanced wireless network simulators capable of realistic modeling of wireless networks, but these simulators are not designed for simulating transportation applications.

While there is agreement among experts on the lack of practical measurements and studies to characterize the 5.8–5.9 GHz DSRC channels, recent large-scale efforts bring about the possibility to simplify models while accounting for the large number of parameters discussed above. The Research Data Exchange (RDE) from US DOT has been created as a transportation data sharing system for archived and real-time data from multiple vehicle probes to support the development, testing, and demonstration of mobility applications and connected vehicle research. In this chapter, we will be particularly interested in the Safety Pilot Model Deployment (SPMD) dataset that provides kinematic, geospatial and connectivity data of approximately 3000 equipped vehicles in Ann Arbor, Michigan. Can such data simplify the problem of realistic modeling of the wireless communication for traffic simulation?

In this chapter we describe the Connected Vehicle Assessment System (CONVAS), which flexibly integrates a traffic simulator with a communication simulator, providing an ideal platform for co-simulating transportation system applications. Its communication models can be tuned based on real-world measurements (e.g. from the Michigan SPMD test bed) in scenarios such as urban, residential and highway traffic. We advocate for representations of real-world wireless channels that capture uncertainty in the existing data and calibrations procedures. The platform can be used to test, validate and assess vehicular communications under a variety of operating traffic and communication conditions and settings.

The layout of the chapter is as follows: section 2 reviews research related to co-simulation of traffic and wireless communications, realistic channel models and current challenges. Section 3 overviews the CONVAS co-simulation platform and its two main components, traffic and communication simulation modules. Section 4 examines existing channel model options and our approach to implementing a parametric stochastic simulation model, called lognormal-Nakagami. Section 5 discusses the real data used and the procedure to tune the aforementioned channel model's parameters for different scenarios, in agreement with the Michigan Safety

Pilot Model Deployment data capture. Section 6 presents the CONVAS implementation of a connected vehicle application, Intelligent Dilemma Zone Avoidance that tackles the yellow traffic light dilemma by using the Signal Phase and Timing (SPaT) messages. Section 7 discusses overall CONVAS results with the application for both the communication and the application components. Finally, section 8 summarizes the main results of this work and future research ideas.

7.2. Related work

As connected vehicle technology advances and new standards are defined by regulatory and standardization organizations, many efforts have been seen in the literature to allow testing of applications using this technology. A common factor in these efforts is the integration between a road traffic simulator and a network simulator, thus attempting to provide realistic models in both scopes. However, the manner in which the interlinking between simulators is achieved can greatly affect the realism with which Connected Vehicle (CV) applications are evaluated.

One approach in interlinking the two simulators is executing the road traffic simulation and exporting its results, i.e. the mobility traces, to the network simulator. Examples of such an approach can be found in [BLU 04, KAI 11]. However, it has been shown in [SOM 08b] that this approach does not model the effects of inter-vehicle communication on the driving behavior. For example, it cannot show how vehicles can change their paths or lanes to avoid congestion based on the traffic information received in the vehicular network. Therefore, it is not suitable for providing an in-depth evaluation on the performance of Connected Vehicle Applications.

Examples of co-simulators which present a finely grained interaction between traffic and network simulators are Veins [SOM 11b], TRANS [PIO 08], NCTUns [WAN 09], iTETRIS [RON 13], MOVES [BON 08] and MSIE [LOC 05]. In particular, Veins supports a coupled network and road traffic simulation using well-established simulators from both communities, namely OMNeT++ and SUMO respectively. The co-simulation is achieved by extending each simulator with a communication module, enabling exchange of commands and mobility traces via TCP connections. From the communication simulation perspective, the OMNeT++ default channel models were extended with propagation models for two-ray interference [SOM 11c] and signal attenuation by buildings [SOM 11a], which were validated on experiments with a few vehicles on different scenarios.

Another co-simulator worth considering is iTETRIS. Its architecture comprises a unique module, the iCS Facilities, interfacing three well-defined environments: (1) iAPP or Applications; (2) ns-3 as a network simulator; (3) SUMO as a traffic simulator. iTETRIS implements four different access technologies in ns-3: ETSI IST G5A, WiMAX, UMTS and DVB-H. In addition, ns-3's default propagation models have been extended to include models fit for urban and highway scenarios taken from the open literature [CHE 07, WIN 07].

However, in spite of allowing for tight interlinking between simulators, the validation of the realism of the presented models against real-world data, if present, was carried out only on a small set of vehicles and road-side equipment. Even in the cases when the authors used widely accepted simulation tools, such as ns-3 or OPNET, there is a need to calibrate and extend the off-the-shelf network and propagation models in order to adapt them to the standards of interest and to the peculiarities of the vehicular environment. In addition to this, the modeling of a particular application is only explicitly visible to the end user in the cases of iTETRIS and MSIE, in which the authors, however, used off-the-shelf versions of the network simulators.

Related areas of research are the evaluation of appropriate mobility models for vehicle-to-vehicle and vehicle-to-infrastructure communication and measurement studies to confirm theoretical characterizations of the DSRC channels. Many of the studies regarding channel models for vehicular communication converge on the use of the Nakagami model, which has been well developed theoretically. In spite of this, there is no agreed upon overall channel model for V2V, and researchers typically adopt combinations of models satisfying various constraints (e.g. short versus large distances, line-of-sight (LOS) versus no line-of-sight (NLOS)). In contrast to the wealth of literature on channel models, there is a lack of data from practical measurements. The many interference opportunities for the 5.9GHz band for DSRC make it even harder to converge on one standard model. Some of these are radar, fixed satellite services, amateur use for the same band, and use of the adjacent bands below 5850 MHz and above 5925 MHz. Industrial, scientific and medical operations of the same band are additional sources of interference in the 5850–5875 MHz portion of the band. This section of the spectrum is presently being evaluated for sharing with WiFi devices based on the IEEE 802.11ac standard [CHA 15].

Table 7.1 summarizes the most relevant state-of-the-art models and parameters applied in vehicular networks. In order to derive the models and parameters, the authors carried out field trials on a small set of vehicles and Road Side Units in different environments. The models account for the signal attenuation over distance by means of the path loss exponent α; occasionally represented as a dual slope model, i.e. different values of α as a function of distance. Large-scale fading and small-scale fading are represented by σ, the standard deviation used in lognormal shadowing, and m, the shape of the gamma distribution in the Nakagami model. In addition, PL_0 represents the additional constant loss added to the total, which is measured at a close distance d_0 to the transmitter.

At the same time, other authors heuristically define or select parameters for their propagation models. This is the case of [HAF 13], in which the authors present an analytic mobility model and its performance analysis for broadcasting via DSRC while taking into account the distance between the transmitter and receiver, speed and vehicle densities using the Nakagami propagation model.

Environment	Model	d_0(m)	PL_0(dB)	α	σ	m	Reference
Urban	Free space	-	-	2.2	-	-	[SOM 11c]
"Free Space" Highway	Free space	-	-	2	-	-	[EEN 09]
Urban LOS	Free space	-	-	2.7–5	-	-	[EEN 09]
Urban NLOS	Free space	-	-	3–5	-	-	[EEN 09]
"Outdoor" Highway	Lognormal shadowing	-	-	2	4–12	-	[EEN 09]
Urban	Log-distance	30	80	2.02–2.13	-	-	[ROI 14]
Suburban	Lognormal shadowing	-	-	2.56	4.0	-	[KAR 07]
Highway	Lognormal shadowing	10	63.3	1.77	3.1	-	[KAR 11]
Urban	Lognormal shadowing	10	62	1.68	1.7	-	[KAR 11]
Suburban	Lognormal shadowing	10	64.6	1.59	2.1	-	[KAR 11]
Suburban	Nakagami	-	-	2.1–3.8	-	0.16–5.8	[ISL 13] Dual Slope model
Suburban	Nakagami	-	-	2.2–2.4	-	1	[BAG 12]

Table 7.1. *Parameter selections for common propagation models*

As stated, the wide variety of propagation models and parameters in the literature is partially caused by a lack of data from extensive field trials covering real-world scenarios. In [DRE 14], the authors list three relevant efforts for real-world data collection: (1) the simTD German project [STU 10], conducted by professional instructed drivers in a controlled environment; (2) the ongoing CAMP (Crash Avoidance Metrics Partnership) project [LUK 12], whose data has not yet been made publicly available; (3) the SPMD (Safety Pilot Model Deployment) [MIC 12], with data readily available from experiments carried out in Michigan. The SPMD stands out in this list as it appears to capture real-world driver behavior with greater accuracy given that the experiments were conducted by common drivers who were allowed to move freely in the city and surroundings of Ann Arbor, Michigan.

In contrast to the multitude of attempts to describe the propagation models for vehicular environments, ITS applications have not been explored in the same measure by the research community. iTETRIS, however, presented two applications in an effort to illustrate the capabilities and potential of the platform. The applications are Cooperative Traffic Congestion Detection (CoTEC) that enables detection of congestion without any fixed infrastructure sensors, and Cooperative Bus Lane Management for allowing private vehicles to use bus lanes when high traffic density is detected.

7.3. CONVAS co-simulation platform

CONVAS is a platform used to test and evaluate connected vehicle applications, automated control, and autonomous driving technologies using simulations of both traffic and communications, which emphasizes the use of commercially available tools, existing communication protocols and standards. Two types of simulators form the skeleton of CONVAS: traffic simulation, such as the Vissim microscopic traffic simulation system, and communication simulation, such as ns-3 or OPNET models and tools. The key feature of CONVAS is the tight integration of the two simulators, in such a way that future events in the traffic simulator are influenced by previous events in the communication simulator and *vice versa*. The resolution of the co-simulation is determined by the configurable parameter Δt, namely, the period of time each simulator runs individually (e.g. 100 msec).

Vissim traffic simulation creates and populates a given traffic scenario consisting of vehicles and infrastructure elements, defines the vehicle control logic and driver behavior parameters for all vehicles within the simulation environment, exchanges traffic control events and status, and ensures vehicle movement in the simulated world that replicates the physics grounded movement of real vehicles. Vissim is a *time-based simulation* system, which will advance simulation time by the constant time step Δt.

Communication simulation is driven by a *discrete event network simulator* that provides high-fidelity packet transmission modeling and detailed analysis capabilities for very large wired and wireless networks. It should allow CV application developers to assess the impact of real-world communication issues such as received power, signal-to-noise ratio, path loss, channel utilization, packet errors and packet delays on the performance of the connected vehicles and infrastructure-based applications. For V2X applications, we are interested in modeling networks using up-to-date wireless protocols such as WiFi-based DSRC, LTE and WiMAX. We have considered two options: OPNET Modeler from Riverbed Communications and ns-3. In the case of OPNET, the wireless models have been extensively tested by consortia. In addition, the open-source alternative ns-3 includes a sufficiently detailed DSRC model that implements the 802.11p and WAVE 1609 protocols.

The details of the simulation, including the application to be simulated, the communication parameters (such as propagation model or transmission power), the traffic density, driver profiles, and other traffic parameters such as Connected Vehicle penetration, are configured in what we call the Application Testing Environment (ATE). Presently, CONVAS has interfaces supporting Vissim for traffic simulation and both OPNET Modeler and ns-3 for network simulation. Details of our implementation are given below.

Initialization of a simulation defines all the static and mobile nodes that could transmit information throughout the entire duration of the simulation. Each node is uniquely identifiable and will be known in both traffic and communication simulators. For each node, information such as the type of node (static or mobile), initial position, antenna characteristics (power, orientation, pattern, etc.) is specified. The traffic simulator passes to the communication simulator: (1) the set of nodes that transmit new information over the most recent Δt period and, for each such node, also the number and type of packet sent and the size of payload (i.e. the amount of information to be transmitted); (2) the present position and the heading over the duration Δt for each mobile node. The trajectory for each mobile node is assumed to be linear and at uniform velocity between two points. The communication simulator passes to the traffic simulator the set of the nodes that will have received packets over the most recent Δt execution period and, for each such node, the type of packet received and the time when the packet was received relative to the present interval.

Their integration follows a client–server model, where the Traffic Simulation Environment (TSE), namely Vissim, is the server and the Communication Simulation Environment (CSE), namely ns-3 or OPNET, acts as a client, and is described in more detail in [SON 17]. Specifically, we implemented TCP sockets directly into both Vissim API and each network simulator. The server will be in listening mode waiting for the socket connection from the CSE upon the start of the simulation. Once the connection is established, each simulation time step is advanced with the exchange of the data through the socket.

7.4. Realistic DSRC channel models

The wireless medium in vehicular communications has characteristics that make it unique. Both transmitters and receivers are mobile and their relative movement creates Doppler shifts, there exist large metal objects constantly moving (other cars), antennas are placed at low elevations and the channel is statistically non-stationary or even random. The strong dependence on the environment and the dynamism of the state of the vehicular environment make it necessary to differentiate among several types of scenarios of interest, such as urban, residential and highway scenarios.

Specifically, we are interested in simulating communication in the following prototypical scenarios in an intersection or on the road: (1) *Urban* scenario represents streets with two to four lanes guarded by large buildings from the side, or most sides of an intersection, and sidewalks; (2) *Residential/Suburban* scenario represents residential two lane streets and intersections in residential areas, characterized by smaller dimensions, possible winding shape, with trees and parceled houses along the street; (3) *Highway* scenario represents broader arteries with two to three lanes per direction, which are flanked by large open spaces and forests.

In simulations of vehicular wireless communications, the propagation model used plays an important role since the received power is crucial when determining whether a packet is received or not. In order to simulate the wireless medium realistically, we distinguish large-scale and small-scale effects in radio wave propagation phenomena. Large-scale effects include reflection, diffraction and scattering. Reflection occurs when a wave encounters a large medium with a different refractive index to air. In models, reflection is often translated to a path loss exponent. Diffraction is a phenomenon explained by Huygens' principle, which states that every point on a wave front acts as a seed for a subsequent wave front to enable waves to propagate around edges or holes. This effect can be modeled with the knife-edge diffraction model, which can be used for site-specific modeling of propagation over hills and large buildings, for example. Scattering of a radio wave occurs when the wave encounters an object whose size is comparable to the wavelength, of the order of tens of centimeters. The effect of this phenomenon is the spreading of the wave in all directions. This can account for a received signal that is stronger than what would have been predicted by reflection or diffraction alone. Scenarios (1) and (3) above have strong large-scale effects that need to be captured by our model.

Small-scale effects include fading. At the receiver, multiple versions of the original signal superimpose. They may be reflected and diffracted, and arrive with time and phase differences. These multi-path waves interfere with each other, which can cause large fluctuations in signal quality with apparently small changes in time or receiver location. This relative motion causes frequency modulation because each multi-path will have a different Doppler shift (variation in the perceived frequency of the signal as a result of the relative speed between the receiver and the transmitter). V2V channels tend to show higher Doppler spreads than conventional mobile radio channels. Scenario (2) exhibits small-scale effects that will be captured by our model.

The OPNET Modeler is equipped with several propagation models: Free space, Longley-Rice, forest, CCIR, HATA and Walfisch–Ikegami [RIV 16]. ns-3 similarly offers the following channel models: Friis, Two-ray ground propagation, Log-distance, Nakagami, Range, etc. [NS 15]. Many of these models are restricted to certain frequencies and distance ranges or are simply not suitable for vehicular environment.

The next section describes how we built on the existing channel models to create and tune our channel model based on real-world data. We first formally describe the channel model used, a combination of lognormal and Nakagami models. We then show how real data is used to tune the parameters of the resulting lognormal-Nakagami channel model.

7.4.1. *CONVAS propagation models*

We have created and tuned specific channel models to extend the standard options available and take into account the special characteristics of the transportation environment. A deterministic model, taking into account signal attenuation by buildings or ray tracing, results in very accurate estimations of path loss and other signal degradation effects. However, its implementation would introduce significant overhead and increase simulation time. The path loss algorithm needs to be executed for every pair of TX and RX antennas on a per packet basis. In the case of Ray tracing, it is necessary to model all rays emanating from the source towards every receiver due to the broadcast nature of the communication. Thus, the number of potential receivers scales up with the square of the number of vehicles times the number of paths. On the other hand, stochastic models offer less accurate but reliable enough results in exchange for a rapid execution and easy implementation. The most widely accepted stochastic model for the simulation of vehicular communications is the Nakagami model. Rician distributions model fading with a single stronger line-of-sight in the presence of scatterers, while Rayleigh distributions are used to model dense scatterers when no line-of-sight is present. The Nakagami distribution is the more general model that can represent Rician, Rayleigh or fading that is more severe than Rayleigh, depending on model parameters, and thus is capable of describing a wide range of fading situations.

We extended the ns-3 WAVE component for vehicular communications with a lognormal-Nakagami propagation model that can account for real-world shadowing and multi-path fading of the wireless signal, known to be predominant effects in the attenuation of the signal in a vehicular environment (see [EEN 09]). The new propagation model is a combination of two already existing models: lognormal shadowing and Nakagami. It calculates the received power in two steps. First (see [MEC 11]) we take into account the effect of shadowing, where the reception power at a distance x to the transmitter is calculated with a log-distance rule as follows:

$$P_r^{LogNorm}(x) = P_t - \left(PL(d_0) + 10\alpha log(\frac{x}{d_0}) + X_\sigma\right)$$

where α is the path loss exponent, $PL(d_0)$ is the path loss measured at a close distance ($x = d_0$) to the transmitter and σ is the standard deviation of the normal and zero mean random variable X representing the Gaussian noise. The power is expressed in dBm and the path loss is expressed in dB. Second, we apply the effects of multi-path fading, modeled by the Nakagami distribution, to obtain the final received signal power:

$$P_r(x,m) = Gamma\Big(m, \frac{P_r^{LogNorm}(x)}{m}\Big)$$

where m controls the shape of the gamma distribution. This formula shows the power expressed in Watts. Depending on the parameter values used in both equations, we can simulate different types of scenarios, such as urban, suburban and highway. Parameter sets for these scenarios have been tuned separately from the real-world data (see the following subsection).

In summary, the set Θ of parameters for this model is given by: (1) the path loss measured at a close distance to the transmitter PL_0; (2) d_0 the reference distance close to the transmitter for PL_0; (3) the path loss exponent α, which varies significantly for different environments; (4) the standard deviation σ of a normal and zero mean random variable X, which represents shadowing fluctuations; (5) the gamma shape of the Nakagami model m, which represents the intensity of the small scale fading effects.

Our lognormal-Nakagami model is therefore parameterized by:

$$\Theta^{Lognorm-Nakagami} = \{PL_0, d_0, \alpha, \sigma, m\}$$

7.4.2. *Model tuning based on real-world data*

The specific combination of parameters Θ significantly affects the resulting reception profile and therefore the CV application performance indicators. The related literature provides many examples of parameter settings for different environments; however, the resulting settings do not always match real-world data. Furthermore, there is a wide variability in the choices for these parameters and also in the specifics of the models. Section 7.2 of this chapter presented a selection of models, scenarios and parameter settings used in the literature. Our approach has been to optimize the model parameters in order to fit real-world measurements (see section 7.5) for the scenarios outlined. By varying the set of parameters, we are able to simulate all these scenarios using one formal model.

Now, we turn our attention to the criterion for tuning the parameters of the lognormal-Nakagami model based on real measurements. The paramount concern in connected vehicle applications is the guarantee for message delivery. A successful application implementation guarantees the communication of sufficient information for closing the vehicle control loop (with the human in or out of the loop) and affecting how vehicles drive. Typical measures used to capture the reliability of

DSRC wireless channel include *average packet delay* (APD), *consecutive packet drops* (CPD) and *packet delivery ratio* (PDR) [BAI 06]. PDR is of interest here and is denoted as $p_r(x)$, being defined as the probability of successfully receiving a packet at a receiver located at a distance x from the sender after broadcast. We will use PDR curves estimated from real-world data for the scenarios of interest over various transmitter–receiver distances x. Our optimization criterion minimizes the integral of the absolute difference between the PDR of the lognormal-Nakagami model M, $p_r^M(x, \Theta^M)$, and the PDR measured from data $p_r^{Real}(x)$ as functions of distance:

$$f(\Theta) = \int_0^R |p_r^M(x, \Theta^M) - p_r^{Real}(x)| dx$$

$$\Theta^* = argmin_\Theta f(\Theta)$$

where R is the transmission range.

We perform global optimization of the parameter set Θ using, for example, simulated annealing (SA), a simple randomized technique for iterative improvement [KIR 83]. SA repeatedly traverses a Markov chain of iteratively improved suboptimal parameter settings by sampling the search space of acceptable settings.

7.5. Channel model tuning

7.5.1. *Michigan safety pilot model deployment data*

CONVAS channel models were tuned based on data collected during the Michigan Safety Pilot Model Deployment (SPMD) [MIC 12], a research initiative that featured real-world implementation of CV safety technologies, applications and systems using everyday drivers. The SPMD data was collected at a test site under real conditions with multi-modal traffic around Ann Arbor, Michigan, from approximately 3000 vehicles equipped with V2V communication devices in an area that spans more than 4000 square miles from Medina in the Southwest corner to Auburn in the Northeast corner.

In the field test, Basic Safety Messages (BSMs) were transmitted at a frequency of 10 Hz. Around 75 percent of the vehicles had transmit-only capabilities, while the rest could transmit, receive and log information at a 10 Hz rate. Vehicles with logging capabilities record two different databases. In short, these had the following information that was used in our analysis (while other available details were not considered in this study): (1) *DAS1-DataRV* recorded entries for every received BSM, its vehicle ID, trip ID, time of reception and position coordinates of the transmitter; (2) *DAS1-DataWSU* recorded the vehicle ID, trip ID, time and position coordinates at a 10 Hz rate.

We selected data corresponding to the three different scenarios: urban, highway and residential/suburban. The selection was done by mapping position coordinates of the vehicles to recognizable geographic areas whose characteristics match those of the scenarios of interest (see Figure 7.1).

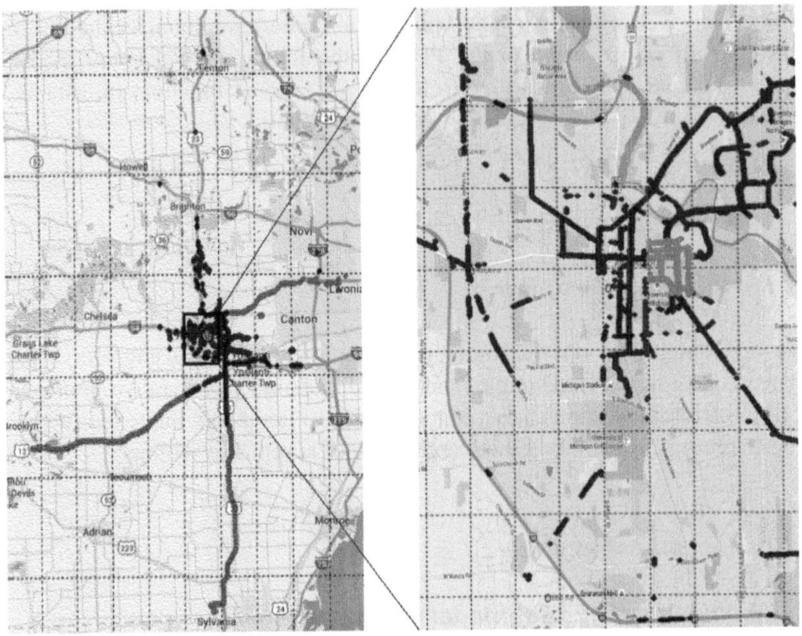

Figure 7.1. *SPMD data collection region in Ann Arbor, Michigan (left), and zoom in for the downtown area (right). Data for the highway, residential/suburban and urban areas selected is marked in red, blue and green, respectively. For a color version of this figure, see www.iste.co.uk/hilt/transportation.zip*

7.5.2. *Estimation of PDR*

Our main goal is to obtain a reference Packet Delivery Ratio (PDR) curve over Tx-Rx distance for each of these scenarios. The data recorded in both databases allows us to compute all distances at which a successful reception occurred. For each distance, we also record the time of reception, the receiver and transmitter IDs, the trip ID and the position coordinates of the receiver.

Instances of distances at which reception failed can only be estimated. For this, we consider each interaction between any two vehicles separately. An *interaction* is defined by the two vehicle IDs and the trip ID of the receiver, where the maximum contiguous gap of packets not received during the interaction cannot be longer than 60 seconds. This time is equivalent to the time it takes to ride along a 200 m street block at a low urban speed average of 10–15 mph and a stop at one traffic light, or a longer residential stretch of road of 500 m at a higher average residential speed of 30–40 mph and a stop at one traffic light. For each interaction, we identify the first and the last packets received and assess lost packets based on the time gaps when no packets are received. Packet receptions are expected at every 100 ms. Whenever the time gap exceeds this value, we estimate distances for lost packets by interpolation based on the distances from the last received packet to the next received packet in periods of 100 ms within the interval of reception. For example, Figure 7.2 depicts eight examples out of the 536 interactions found in the urban region. Note that in many of these cases the two vehicles begin interacting at larger distances, only to converge and further depart away such as when two vehicles travel in opposite directions.

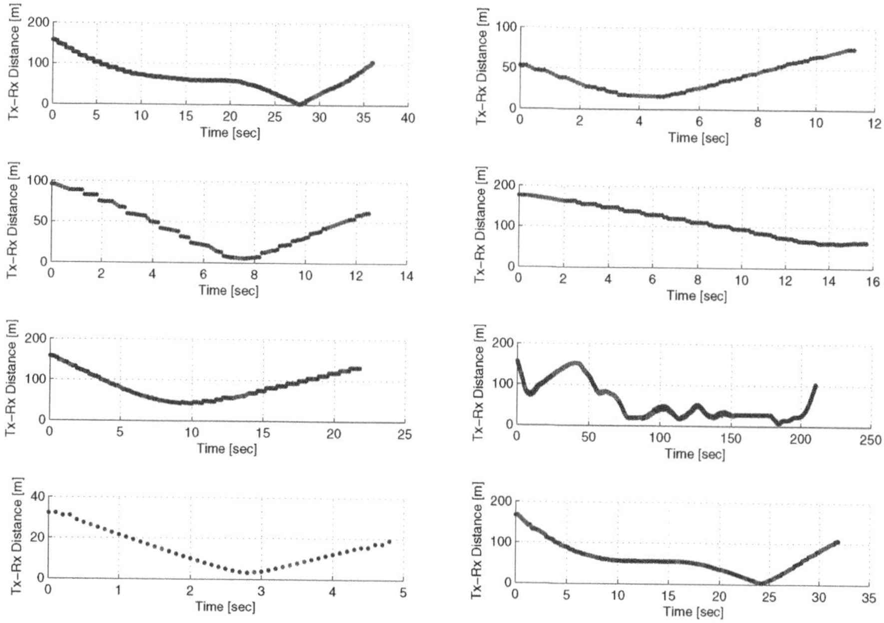

Figure 7.2. *Examples of Tx-Rx interactions found in the urban region for the SPMD dataset, where received packets are blue and estimated lost packets are red. This allows the estimation of interaction distances for lost packets. For a color version of this figure, see www.iste.co.uk/hilt/transportation.zip*

As a result of applying the previously described procedure, we obtain a distribution of received packets and a lower bound of the number of lost packets for each one of the regions of interest. We consider only the distances in the range from 0 to 500 meters, since the typical transmission range for DSRC radios is

R = 250...300 meters[1] [QIN 04, SHO 09]. Given this and the definition of a wireless communication interaction between two vehicles given earlier, the estimation of channel models will be also considered over lengths of 250 m for urban and 500 m for residential scenarios. Figure 7.3 shows the distributions obtained for urban and residential/suburban environments with 536 and 1123 interactions, for a total number of 128,162 and 237,395 packets received respectively. The selected highway regions comprised only several interactions, and therefore do not offer sufficient statistics for further analysis of the real highway environment.

Note that the number of lost packets for distances approaching the transmission range decreases with distance. Intuitively this number should increase; however, the number of packets lost at larger distances is likely to be heavily underestimated due to the computational assumption that there exist no (lost) packets before and after the first received packet for any interaction. This typically happens for large distances rather than for short distances. The confidence in the $p_r^{Real}(x)$ estimates is high for short distances, where there exists more than 300,000 packets received at such distances, and decreases when the distance approaches the transmission range, where the number of packets received is less than 0.5% of the corresponding number for short distances. These statistics from the real world also suggest that the estimated PDR, $p_r^{Real}(x)$, will be accurate only for short distances given the amount of data in SPMD.

Finally, we obtain the PDR as a function of distance, given by:

$$p_r^{Real}(x) = \frac{N_r(x)}{N_r(x) + \hat{N}_l(x)}$$

where $N_r(x)$ and $\hat{N}_l(x)$ are the number of packets received and the estimate of the number of packets lost respectively, at a distance x from the sender. $p_r^{Real}(x)$ represents an optimistic packet reception rate in the real world that is close to the true rate at the receiver for short distances. Figure 7.4 represents PDR for the urban and residential/suburban scenarios where a reasonable amount of logged data was available in SPMD.

7.5.3. *Model tuning*

We match specific real-world scenarios as defined by the Safety Pilot test bed data [MIC 12]. The optimal parameters defining each scenario were initialized based on other studies from the literature and further optimized by simulated annealing. Note that we used a fixed value of 10 meters for the reference distance d_0 and we assumed that the SPMD transmission power was 23 dBm and the reception power threshold −88 dBm. Figure 7.4 shows the best solutions obtained after global optimization of the parameters as above.

1 In the literature, the commonly cited transmission power level is 23 dBm (0.1995 W). The maximum allowed power level by US FCC for the 5.9 GHz DSRC band is 33 dBm.

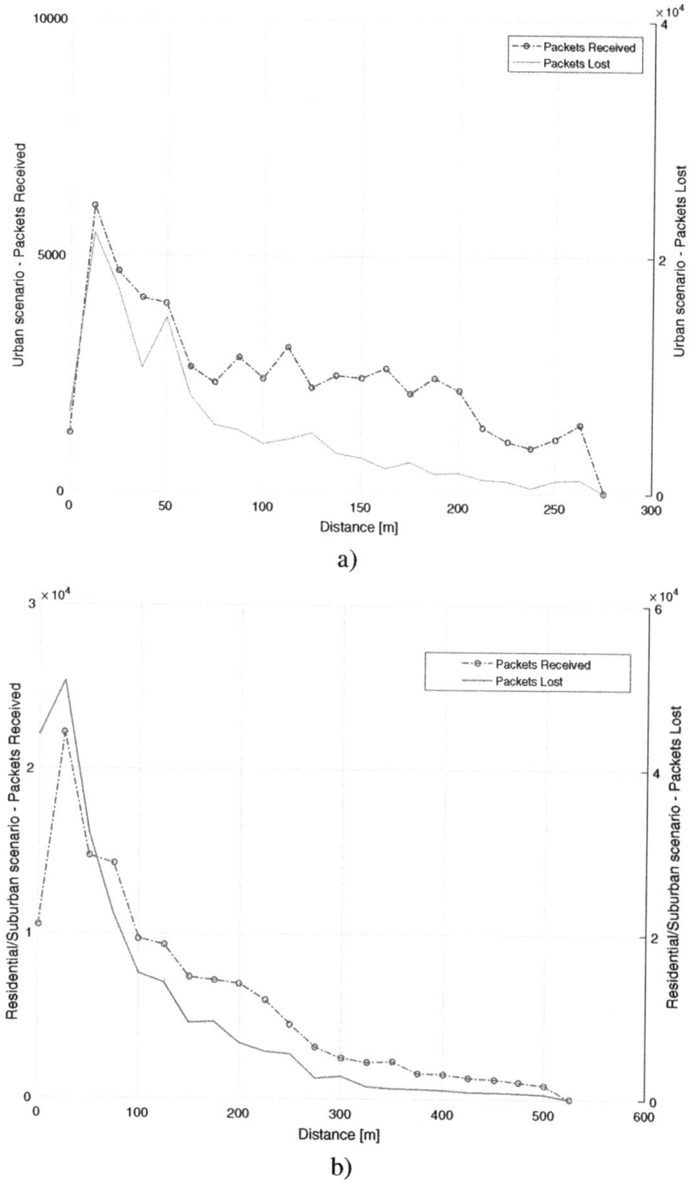

Figure 7.3. *Received and lost packet distributions as functions of distance for: a) urban environment; b) residential/suburban environment in the SPMD data. For a color version of this figure, see www.iste.co.uk/hilt/transportation.zip*

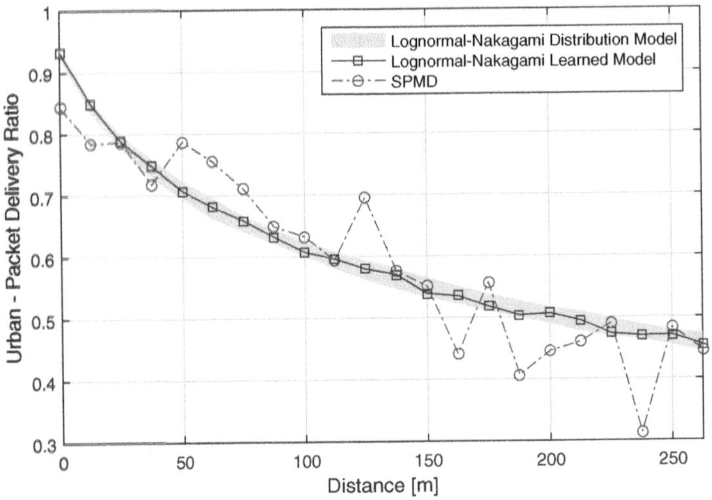

a)

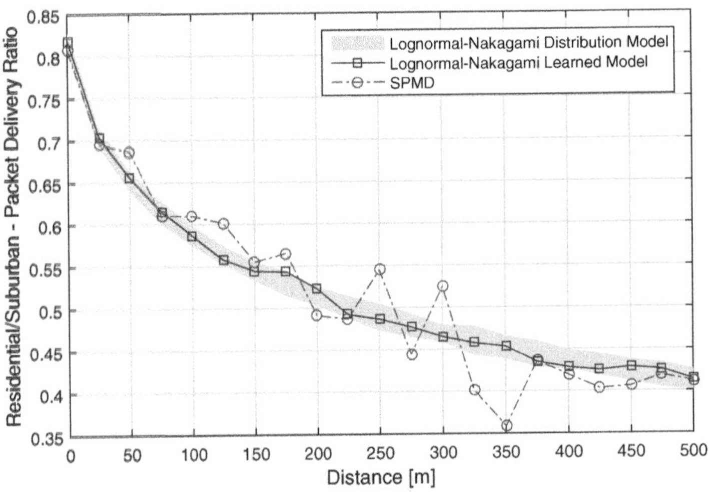

b)

Figure 7.4. *Estimated Packet Delivery Rate (PDR): (a) urban environment; (b) residential/suburban environment. For a color version of this figure, see www.iste.co.uk/hilt/transportation.zip*

The corresponding values of the parameters Θ presented in Table 7.2 for the scenarios of interest have been used further in the ns-3 lognormal-Nakagami propagation model for co-simulation experiments with various applications in the urban, residential/suburban or highway scenarios. When compared to the parameters commonly used in the literature in Table 7.1, we observe that our values for PL_0 and σ are much higher, thus showing that, according to the SPMD dataset, propagation conditions are much worse than previously studied. Also, $m > 1$, therefore the shape of the gamma distribution indicates Rician fading for both scenarios, which suggest dominant line-of-sight settings. Overall, this means that in the residential/suburban scenario the signal presents a mix of propagation along a strong line of sight and non LOS between transmitter and receiver.

Environment	d_0 $[m]$	PL_0 $[dB]$	α	σ	m
Urban	10	89.96	1.50	15.26	2.06
Residential/Suburban	10	94.97	1.08	16.05	2.14
Highway	10	63.3	1.77	3.1	-

Table 7.2. *Parameters for Lognormal-Nakagami model tuned to match urban and residential/suburban scenarios from the SPMD dataset, and [KAR 11] for highway scenarios*

Figure 7.4 also presents, for each of the urban and residential/suburban scenarios, the lognormal-Nakagami model learned from SPMD data, and a stochastic model given by a collection of probability mass functions, one for each distance interval, called the distribution model. The latter has been obtained from Monte Carlo simulations using the parameters of the learned lognormal-Nakagami model. The distribution model can be updated from new real world measurements and plugged into the communication simulation. It better captures the uncertainty in the real-world data used for training.

7.6. Connected vehicle applications

The quality of the wireless communications will affect the performance of large-scale connected and automated vehicle (CV/AV) applications. This can be studied in CONVAS without the effort and costs of real-world deployments. Examples of applications being modeled using CONVAS include Forward Collision Warning, Cooperative Adaptive Cruise Control and Intelligent Signalized Intersection Control. In this chapter, we describe and present results with an application called Intelligent Dilemma Zone Avoidance (IDZA) [SON 17] to demonstrate the use of the CONVAS platform for testing applications.

7.6.1. *Intelligent dilemma zone avoidance*

The goal of this application is to provide automated warnings to the driver or even control the vehicle automatically in order to solve the dilemma of whether to slam on the brakes and risk being rear-ended, or speed through the light and risk a collision or a traffic ticket while driving towards an intersection when the traffic light changes from green to yellow. The distance around the traffic light where this decision must be made is called dilemma zone (DZ). Certainly, the dilemma zone depends on the driving speed and the geometrical configuration of the intersection.

Technically, IDZA aims to provide automated longitudinal control to the connected and automated (equipped) vehicles to avoid getting trapped in the dilemma zone by using current vehicle state and Signal Phase and Timing (SPaT) messages from signalized intersections. The application resides within the vehicle on-board unit (OBU) and will engage the vehicle's throttle if it is predicted to be in the DZ based on its instantaneous speed upon the reception of the SPaT message.

7.6.2. *IDZA implementation in CONVAS*

Details of the modeling and implementation of IDZA are presented in [SON 17]. The IDZA application is modeled within a Vissim environment and relies on a road side equipment unit located in a corner of the signalized intersection and capable of broadcasting SPaT messages. The wireless communication model simulates the message delay and reception every time step for all equipped vehicles in the network. Upon the reception of the message, vehicles determine if they are predicted to be in the DZ and dynamically adjust acceleration/deceleration rates as necessary to prevent themselves from getting trapped in the DZ, while the lead vehicle continues at its current speed. The application can take control of the vehicle's throttle within 100 ms. The model includes a delay for switching from manual to automated driving upon the detection of potential DZ traps.

The application implements the following conditions for the activation (manual to automated) of IDZA: (1) the vehicle speed has to be higher than the minimum speed threshold; (2) the vehicle must be approaching and no more than the minimum time to stop bar threshold; (3) the vehicle has to approach the green indication of the phases designated for DZ avoidance. The conditions for deactivation (automated to manual) are: (1) acceleration required is outside the applicable range or (2) the vehicle loses the reception of SPaT messages.

Furthermore, during the auto-to-manual transition, the vehicle control uses the acceleration of the previous time step. If a vehicle has been in accelerating mode, the control reverts to the cruise mode, with zero acceleration. If a vehicle has been in decelerating mode, the control continues to decelerate at the same rate during the transition. The automatic to manual transition is canceled if the automated longitudinal control starts within the configured period. If all the conditions are met and both acceleration and deceleration options are viable, the algorithm chooses the closer edge of the dilemma zone based on its current estimate of time to stop bar at the onset of yellow.

7.6.3. *IDZA performance criteria*

When the vehicle is predicted to be in the DZ, IDZA uses the vehicle kinematics to derive the acceleration required to reach the first or the second edge of the DZ, and thus computes the acceleration rate needed to clear the dilemma zone trap as a function of the phase time remaining to the onset of yellow (from its SPaT message), the vehicle speed and the distance to stop bar (from the radar sensor). A vehicle can then either accelerate to reach 2.5 seconds or decelerate to reach 5.5 seconds as the time to stop bar at the onset of yellow.

Possible performance criteria are: (1) the rate of vehicles trapped in the dilemma zone; (2) the distribution of time to stop; (3) average reception loss from the moment of the first reception. These measures will be tracked in the next section.

7.7. Experimental results

7.7.1. *CONVAS setup*

The objective of the co-simulation experiment is to demonstrate the use of CONVAS and to evaluate the effect of wireless communications on the application performance. The CONVAS initialization sets up CSE (i.e. OPNET or ns-3), TSE (Vissim) and the application parameters. Table 7.3 presents the setup parameters for the wireless simulation (either of OPNET or ns-3).

Parameter	Example	Units	Description
Packet Generation Window	10	ms	Maximum time uniformly distributed [0, window] to wait before generating a new application layer packet
Rx/Tx Additional Delay	15	ms	Processing delay added to the packet in Rx/Tx application layers
Data Rate	12e6	Mbps	MAC layer data rate
Tx Power	0.199	W	PHY transmit power
Rx Power Threshold	–88	dBm	PHY power sensitivity
Min Frequency	5855	MHz	Channel's minimum frequency
Bandwidth	10	MHz	Channel bandwidth
Modulation	OFDM	–	Modulation scheme
Packet Length	40	Bytes	Standard length of BSM Part I

Table 7.3. *Wireless communication simulation parameters*

The following parameters define the IDZA application setup:

– delay in transitioning from DZ automation to manual driving is 1 second (when deactivation condition is satisfied);

– minimum speed threshold is 35 mph;

– minimum time to stop bar for tracking is 10 seconds;

– dilemma zone time thresholds are 2.5 or 5.5 seconds.

Accordingly, the application takes control from the driver only when the following conditions are met:

– the vehicle is approaching a green phase designated for DZ avoidance;

– the vehicle has received the SPaT message;

– the vehicle speed is greater than 35 mph;

– estimated time to stop bar is less than 10 seconds;

– the computed acceleration rate is within comfortable acceleration and deceleration limits of $\pm 2m/s^2$;

– the computed acceleration rate is less than the speed-dependent acceleration rate. The acceleration performance of the vehicle is also known to decrease with vehicle speed. We adopted a linearly decreasing acceleration profile to calculate maximum allowable speed-dependent acceleration rate using the equations reported from [LON 00];

– the front gap corresponds to a time headway currently of minimum 2 seconds.

We set up an isolated signalized intersection test bed in Vissim with an 8-phase fixed time operation. The operating speed is set at 55 mph on all approaches. The IDZA is set to be active on all through phases. Figure 7.5 shows the geometry and layout of the intersection.

7.7.2. *Co-simulation results*

We performed experiments in scenarios with an intersection volume of 2500 vehicles per hour, 75% of the traffic volume being on the major street, and with either 25% or 75% of DSRC equipped vehicles, under various communication conditions ranging from ideal communication to several cases of transmission range, power and precision of the communication simulation, and without or with IDZA control. During a co-simulation run, we log both network statistics and the trajectory of the equipped vehicle from the moment at which it meets the criteria for activating the control until the onset of yellow. This allows the visualization of a dashboard of communications and application statistics. We present sample results next.

Figure 7.5. *Signalized intersection for IDZA. Black vehicles are not DSRC equipped. Yellow vehicles have no reception and are manually driven. Red vehicles have reception and are automatically decelerated. Blue vehicles have reception but are manually driven. Light blue vehicles have reception and are controlled to accelerate. For a color version of this figure, see www.iste.co.uk/hilt/transportation.zip*

Figure 7.6 shows the received power scatter plot for both urban and residential/suburban scenarios. It can be appreciated how the residential/suburban received power is much more compact than in the urban case; this is the outcome of the combined effects of each of the parameters in our lognormal-Nakagami model. For instance, higher values of σ produce more spread received power.

Figure 7.7 shows the number of vehicles present in one instance of the simulation over time and the number of packets received in the network over time. Note an obvious correlation between the two plots.

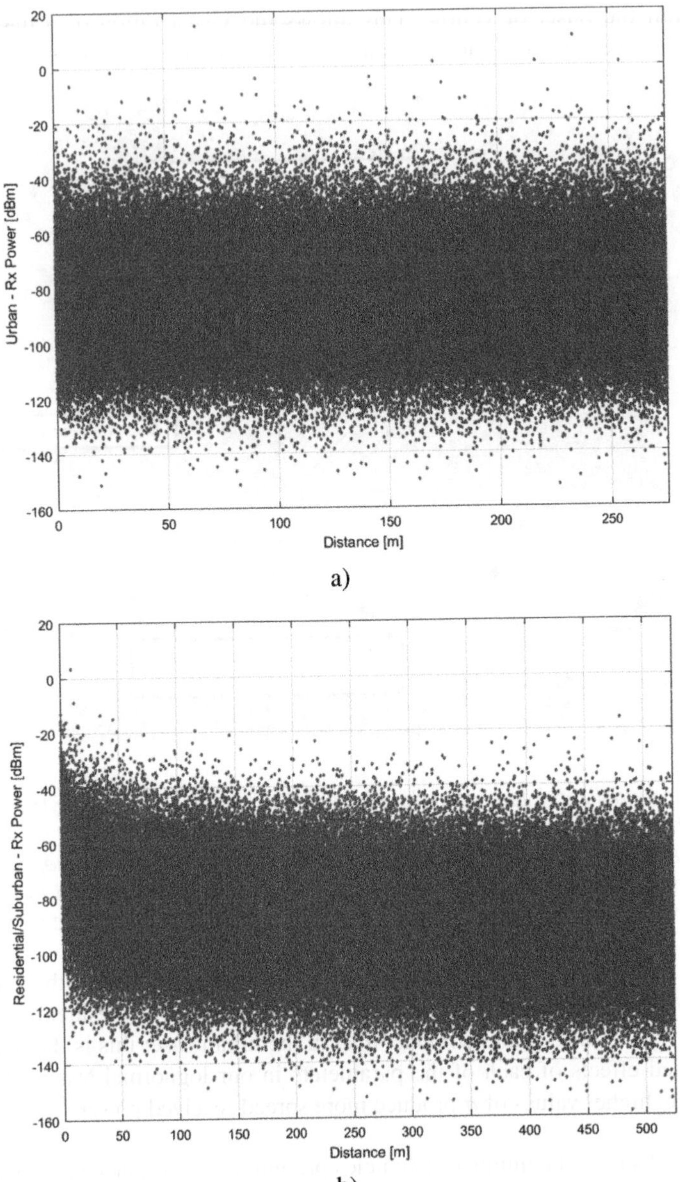

Figure 7.6. *Received power for the lognormal-Nakagami model and the receiver sensitivity (marked in red): a) urban environment; b) residential/suburban environment. For a color version of this figure, see www.iste.co.uk/hilt/transportation.zip*

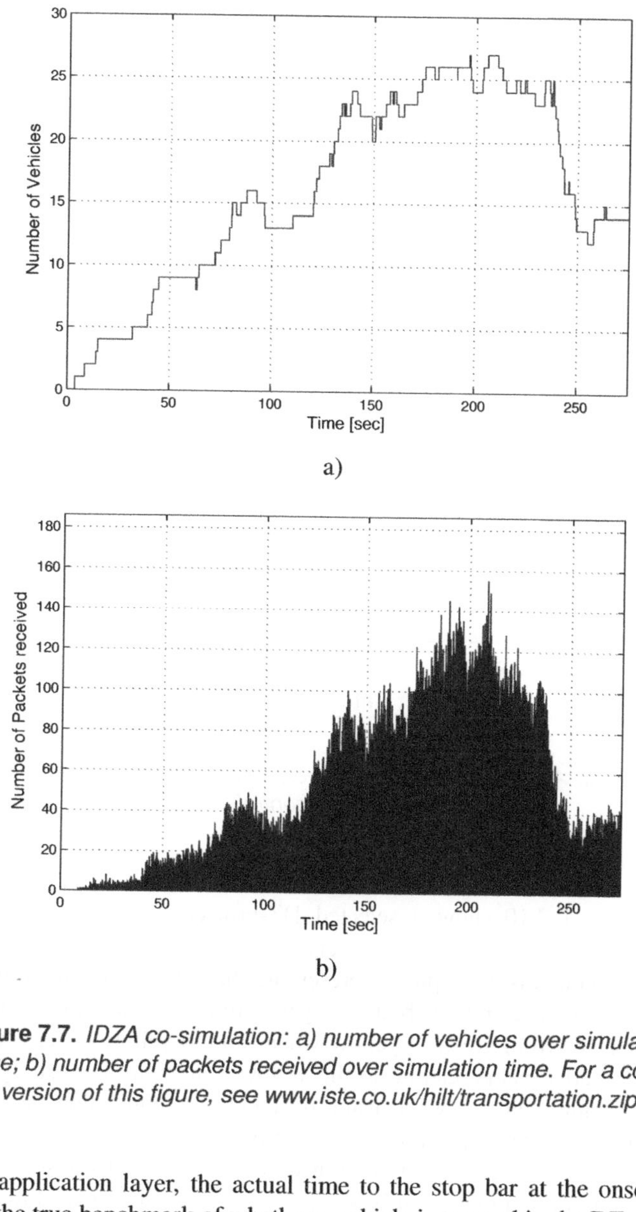

Figure 7.7. *IDZA co-simulation: a) number of vehicles over simulation time; b) number of packets received over simulation time. For a color version of this figure, see www.iste.co.uk/hilt/transportation.zip*

At the application layer, the actual time to the stop bar at the onset of yellow represents the true benchmark of whether a vehicle is trapped in the DZ. Several data attributes are also logged during the approach including vehicle speed, approaching phase, phase indication, actual acceleration, planned acceleration, front gap and SPaT reception status. The log data from CONVAS was used in order to assess the overall performance of IDZA control, given by the trap rate in the danger zone. We experimented with a variety of conditions, representing various penetration rates of DSRC (e.g. 25–30% or 70–75%), volumes of traffic (e.g. 5000 vehicles per hour entering the intersection), communication models (described later in this section), and environments (urban, residential, etc.). The results are consistent across models and show a critical dependence of the performance of the application on reception rates for wireless communication. The results aggregating many hours of co-simulation over tens of experiments are presented in Figure 7.8.

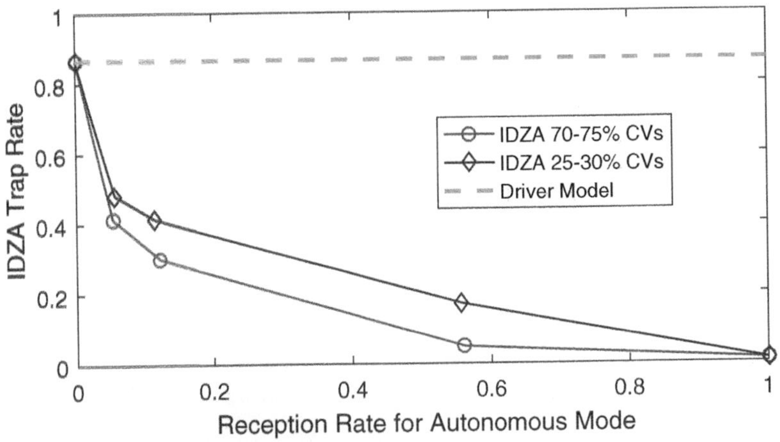

Figure 7.8. *Average Trap Rate vs. Average Reception Rate for IDZA autonomous control*

Figures 7.9 and 7.10 show a selected DZ-projected vehicle's statistics as it approaches the stop bar in one successful and one unsuccessful approach to clear the DZ. Each figure contains three plots representing the vehicle time to stop bar (top), speed (middle), and acceleration (bottom) against time to yellow on the x-axis. The traffic light changes from green to yellow when time to yellow is zero, while prior times have a negative value leading to the yellow transition. The family of curves represent the sample vehicles when using different communication models and an automated control strategy, as follows: (a) Driver Model is the mode where the vehicle has its DZ control algorithm turned off; (b, c, d) ns-3 suburban/urban/highway represent three Lognormal-Nakagami scenarios with channel models tuned based on real measurements. If the projected *Time to stop bar* (top) at $Time\ to\ Yellow = 0$ is outside the [2.5, 5.5] seconds interval, the vehicle will have successfully avoided DZ.

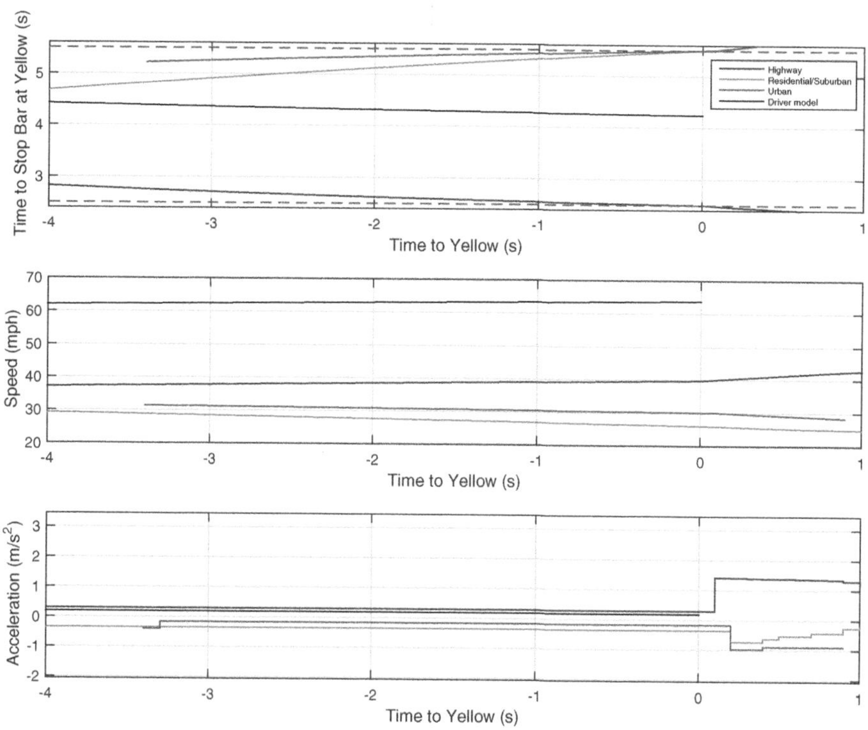

Figure 7.9. *Example of successful automated DZ avoidance under various communication models. For a color version of this figure, see www.iste.co.uk/hilt/transportation.zip*

For example, Figure 7.9 (middle) shows that the vehicle maintains its 65 mph speed in the driver model case to eventually get locked in the DZ, while in all the other cases automated control based on DSRC and radar inputs eventually drive the vehicles faster or slower (with the acceleration as function of time showed in the bottom plot) in order to avoid the DZ (as seen in the top plot). In particular, the vehicle running the Residential/Suburban model breaks and has at least 5.5 seconds projected time to the stop bar when the light turns from green to yellow at time zero. In the example from Figure 7.10 the vehicle control fails to clear the vehicle from the DZ in all cases because the timing of DZ prediction for control activation and the intermittent reception loss during the approach do not allow sufficient time necessary for the algorithm to control the vehicle. The variation in vehicle trajectories under various communications conditions emphasizes the importance of the effect of wireless communications on the automated control outcome.

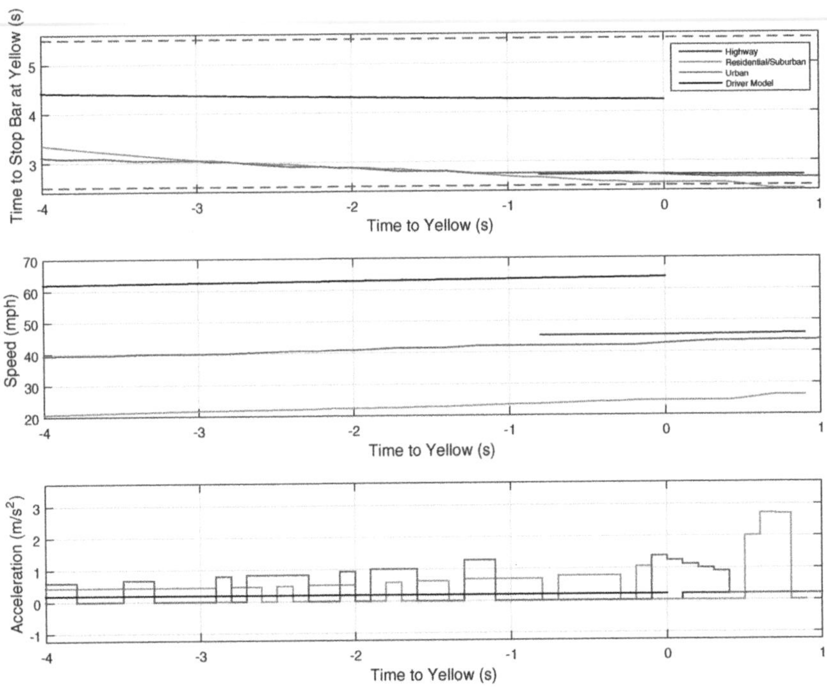

Figure 7.10. *Example of DZ avoidance failure. For a color version of this figure, see www.iste.co.uk/hilt/transportation.zip*

Table 7.4 summarizes some key performance measures collected for each simulation scenario, while a full statistical analysis is undergoing. Note that the default driver model used by VISSIM to control vehicles is the Wiedermann model. In contrast, our IDZA autonomous driving model relies on wireless communication of SPaT messages from the infrastructure. The Nakagami model is a simplified version of wireless communication model where the reception probability depends only on two key parameters, distance and transmission range, for a fixed gamma shape $m = 3$. Nakagami 500 and 1000 represent the case of the plain Nakagami channel model with a transmission range of either 500 or 1000 ft respectively [KIL 09]. Our addition to the ns-3 model is more complex and realistic. It captures factors such as shadowing and multi-path fading of the wireless signal. In addition, it implements the full WAVE protocol stack and offers realistic information about reception time of the packets, which is important for implementing safety applications. In contrast, the Nakagami does not model the WAVE protocol stack, does not indicate reception time and only confirms reception of the packets.

DSRC Vehicles [%]	Control	Comm. Model	IDZA Reception Rate [%]	IDZA Trap Rate [%]
25 – 30	Driver Model	–	–	88.2
25 – 30	IDZA	Nakagami 1000ft	86.3	23.5
25 – 30	IDZA	Nakagami 500ft	37.2	33.3
25 – 30	IDZA	ns-3 Residential	41.0	11.0
25 – 30	IDZA	ns-3 Urban	48.2	5.6
70 – 75	Driver Model	–	–	85.6
70 – 75	IDZA	Nakagami 1000ft	85.6	42.5
70 – 75	IDZA	Nakagami 500ft	42.5	42.5
70 – 75	IDZA	ns-3 Suburban	30.1	12.2
70 – 75	IDZA	ns-3 Urban	41.4	5.3

Table 7.4. *Summary of IDZA simulation scenarios and performance*

7.8. Conclusions

ITS have at their core *connected* vehicles talking to each other using wireless communications. Also, non-vehicle devices such as smart phones, backpacks and bicycles could incorporate the talking technology to communicate with vehicles. These technologies are converging towards real-world use under the big promise of increasing safety for people and vehicles, and improving traffic flow by aiding modern traffic management and autonomous driving. Large-scale simulations are of tremendous value for understanding the critical factors in outstanding safety and mobility applications.

In this work we guide V2V and I2V communication simulation based on *received packet rates*, which is the most critical factor in evaluating a CV application's success. Our goal was to use the largest available dataset showing reception in real environments, from the Michigan test bed. We evaluated applications using propagation settings grounded in the real-world macro-level measurements, i.e. at the level of actually received packets rather than signal power, interference or other low level measures. The results offer a view surprising to the present understanding (from both theoretical and pragmatic perspective) of vehicular channel models. For example, the 50–60% reception rates at short distances in urban areas are factual, and the information then leads to the tuning of the propagation model parameters that are far different from settings seen in previous research. Similarly, our models do not explicitly include a congestion component in packet loss, but rather implicitly account for the effect of high densities of vehicles in the urban scenario. Our results are not as optimistic as other tests under real-world conditions. For example, in [AHM 14] the NHTSA CAMP partnership illustrated an effective average PER below 10% for distances where vehicles spent most of their time. This translated into a PDR of 90%, much higher than the PDRs obtained from the SPMD data. While the CAMP researchers used the same 10 Hz transmission scheme of BSMs, the experiments were conducted in a reduced set of only eight vehicles, which kept the same convoy formation at all times, and the presented PER was a result of the superposition of results obtained in all the tested scenarios: mountainous, deep urban, freeway and major/local roads.

Communication characteristics and quality will have significant impacts on the performance of connected and automated vehicle applications, particularly when implementing safety features as shown in our CV application. For instance, reception of packets affects the timeliness of context information, i.e. awareness of the presence of other vehicles, traffic light status, etc., being made available to the running application. Therefore, realistic communication simulation using models that capture stochasticity as presented in this chapter plays a paramount role in the accurate modeling of connected and automated vehicle technologies. Multiple co-simulators for VANETs can be found in the literature; however, not all of them present an active exchange of information and commands among their components as simulation time advances. CONVAS is a deeply interlinked co-simulator that offers a high level of realism to assist researchers in testing of CV applications. Realism is achieved first by integrating up-to-date communication simulation tools with accurately parameterized propagation models based on real-world measurements, and second by employing a traffic simulation tool with the ability to implement specific vehicle behaviors in order to enable modeling of varied applications.

At the physical level, knowledge has advanced to better understand interference, fading and congestion. However, lack of extensive real-world measurements and usage of various hardware and software settings for the existing measurements lead to inconclusive data sets. The divergence of results from various real-world based studies as shown here highlights the need for rigorous validation of the data and measurements acquired in the community. This research can be extended to consider actuated or adaptive signal control, where the dilemma zone avoidance may be in tighter correlation with safety. An additional direction of future work is the validation of a large number of CV applications in CONVAS, while channel models can be learned and updated as new real-world measurements become available. It will be desirable to model congestion of the wireless medium as bandwidth utilization increases for large penetrations. Awareness of channel utilization can be arguably taken advantage of across layers; however, we do not presently have measurements for high utilization factors. Finally, it is of interest to optimize the CONVAS simulation performance for large-scale traffic scenarios.

7.9. Acknowledgments

This work was funded by the Exploratory Advanced Research Program (EARP) of the Federal Highway Administration (FHWA) – Grant Number DTFH6114C00003. We would like to thank Juan Aparicio for early work on propagation models and discussions in this area, and to Apoorba Bibek for co-simulation testing.

7.10. Bibliography

[AHM 14] AHMED-ZAID F., KRISHNAN H., VLADIMEROU V. *et al.*, Vehicle-to-vehicle safety system and vehicle build for safety pilot (V2V-SP), Final Report Volume 2 of 2, Performance Testing – DRAFT, Crash Avoidance Metrics Partnership, NHTSA, 2014.

[BAG 12] BAGUENA M., CALAFATE C.T., CANO J. *et al.*, "Towards realistic vehicular network simulation models", *Wireless Days (WD)*, pp. 1–3, 2012.

[BAI 06] BAI F., KRISHNAN H., "Reliability analysis of DSRC wireless communication for vehicle safety applications", *IEEE Intelligent Transportation Systems Conference*, pp. 355–362, 2006.

[BLU 04] BLUM J., ESKANDARIAN A., HOFFMAN L., "Challenges of intervehicle ad hoc networks", *IEEE Transactions on Intelligent Transportation Systems*, vol. 5, no. 4, pp. 347–351, 2004.

[BON 02] BONNESON J., MIDDLETON D., ZIMMERMAN K. *et al.*, Intelligent detection-control system for rural signalized intersections, Technical Report FHWA/TX-03/4022-2, Texas Transportation Institute, August 2002.

[BON 08] BONONI L., DI FELICE M., D'ANGELO G. *et al.*, "MoVES: a framework for parallel and distributed simulation of wireless vehicular ad hoc networks", *Computer Networks*, vol. 52, no. 1, pp. 155–179, 2008.

[CHA 15] CHACHICH A., FESSMANN V., ARNOLD J. *et al.*, "DSRC-Unlicensed Device Test plan, USDOT Intelligent Transportation Systems – Joint Program Office", available at: http://www.its.dot.gov/connected_vehicle/pdf/DSRC_TestPlanv3.5.3.pdf, 2015.

[CHE 07] CHENG L., HENTY B.E., STANCIL D.D. *et al.*, "Mobile vehicle-to-vehicle narrow-band channel measurement and characterization of the 5.9 GHz dedicated short range communication (DSRC) frequency band", *IEEE Journal on Selected Areas in Communications*, vol. 25, no. 8, pp. 1501–1516, 2007.

[DRE 14] DRESSLER F., HARTENSTEIN H., ALTINTAS O. *et al.*, "Inter-vehicle communication: Quo vadis", *IEEE Communications Magazine*, vol. 52, no. 6, pp. 170–177, 2014.

[EEN 09] EENENNAAM E.M.V., A Survey of Propagation Models Used in Vehicular Ad Hoc Network (VANET) Research, Faculty of EEMCS, University of Twente, 2009.

[HAF 13] HAFEEZ K.A., ZHAO L., MA B. *et al.*, "Performance analysis and enhancement of the DSRC for VANET's safety applications", *IEEE Transactions on Vehicular Technology*, vol. 62, no. 7, pp. 3069–3083, 2013.

[HWU 88] HWUANG C.R., "Simulated annealing: theory and applications", *Acta Applicandae Mathematicae*, vol. 12, no. 1, pp. 108–111, 1988.

[ISL 13] ISLAM T., HU Y., ONUR E. *et al.*, "Realistic simulation of IEEE 802.11 p channel in mobile vehicle to vehicle communication", *Microwave Techniques (COMITE) Conference*, pp. 156–161, 2013.

[KAI 11] KAISSER F., GRANSART C., KASSAB M. *et al.*, A Framework to Simulate VANET Scenarios with SUMO, University Lille Nord de France, 2011.

[KAR 07] KARNADI F.K., MO Z.H., LAN K.C., "Rapid generation of realistic mobility models for VANET", *Wireless Communications and Networking Conference*, pp. 2506–2511, 2007.

[KAR 11] KAREDAL J., CZINK N., PAIER A. *et al.*, "Path loss modeling for vehicle-to-vehicle communications", *Vehicular Technology, IEEE Transactions*, pp. 323–328, 2011.

[KEN 11] KENNEY J.B., "Dedicated Short-Range Communications (DSRC) standards in the United States",*Proceedings of the IEEE*, vol. 99, no. 7, pp. 1162–1182, 2011.

[KIL 09] KILLAT M., HARTENSTEIN H., "An empirical model for probability of packet reception in vehicular ad hoc networks", *EURASIP Journal on Wireless Communications and Networking*, vol. 2009, no. 1, p. 721301, 2009.

[KIR 83] KIRKPATRICK S., GELATT C.D., VECCHI M.P., "Optimization by simulated annealing", *Science*, vol. 220, no. 4598, pp. 671–680, 1983.

[LEE 15] LEE J., PARK B.B., "Investigating communications performance for automated vehicle-based intersection control under connected vehicle environment", *IEEE Intelligent Vehicles Symposium (IV)*, Seoul, 2015.

[LOC 05] LOCHERT C., BARTHELS A., CERVANTES A. *et al.*, "Multiple simulator interlinking environment for IVC", *2nd ACM International Workshop on Vehicular Ad Hoc Networks (VANET 2005)*, Cologne, pp. 87–88, 2005.

[LON 00] LONG G., "Acceleration characteristics of starting vehicles", *Transportation Research Record*, vol. 1737, pp. 58–70, 2000.

[LUK 12] LUKUC M., *V2V Interoperability Project*, US DOT ITS Connected Vehicle Workshop, Chicago, September 2012.

[MEC 11] MECKLENBRAUKER C.F., MOLISCH A.F., KAREDAL J. *et al.*, "Vehicular channel characterization and its implications for wireless system design and performance", *Proceedings of the IEEE*, vol. 99, pp. 1189–1212, 2011.

[MIC 12] MICHIGAN SAFETY PILOT MODEL DEPLOYMENT, available at: https://www.its-rde.net/data/showds?dataEnvironmentNumber=10018, 2012.

[NS 15] NS-3 MODEL LIBRARY, Release ns-3.24 (September 2015), available at: https://www.nsnam.org/docs/models/ns-3-model-library.pdf, 2015.

[NS 16] NS-3 NETWORK SIMULATOR, available at: https://www.nsnam.org/, accessed on 25 February 2016.

[PIO 08] PIORKOWSKI M., RAYA M., LUGO A. *et al.*, "TraNS: realistic joint traffic and network simulator for VANETs", *ACM SIGMOBILE Mobile Computing and Communications Review*, vol. 12, no. 1, pp. 31–33, 2008.

[QIN 04] QING X., MAK T., KO J. *et al.*, "Vehicle-to-vehicle safety messaging in DSRC", *Proceedings of the 1st ACM International Workshop on Vehicular Ad Hoc Networks*, pp. 19–28, 2004.

[RIV 16] RIVERBED (OPNET) MODELER, available at: http://www.riverbed.com/products/steelcentral/steelcentral-riverbed-modeler.html, 2016.

[ROI 14] ROIVAINEN A., JAYASINGHE P., MEINILA J. *et al.*, "Vehicle-to-vehicle radio channel characterization in urban environment at 2.3 GHz and 5.25 GHz", *IEEE 25th Annual International Symposium on Personal, Indoor, and Mobile Radio Communication (PIMRC)*, pp. 63–67, 2014.

[RON 13] RONDINONE M., MANEROS J., KRAJZEWICZ D. *et al.*, "iTETRIS: a modular simulation platform for the large scale evaluation of cooperative ITS applications", *Simulation Modelling Practice and Theory*, vol. 34, 2013.

[SHO 09] SHOREY R., WEIMERSKIRCH A., JIANG D. *et al.*, "Characterization of DSRC performance as a function of transmit power", *Proceedings of the Sixth International Workshop on Vehicular Ad Hoc Networks (VANET)*, Beijing, ACM, pp. 63–68, 2009.

[SOM 08a] SOMMER C., YAO Z., GERMAN R. *et al.*, "Simulating the influence of IVC on road traffic using bidirectionally coupled simulators", *IEEE INFOCOM Workshops 2008*, Phoenix, pp. 1–6, 2008.

[SOM 08b] SOMMER C., YAO Z., GERMAN R. *et al.*, "On the need for bidirectional coupling of road traffic microsimulation and network simulation", *Proceedings of 9th ACM International Symposium on Mobile Ad Hoc Networking and Computing (Mobihoc 2008): 1st ACM International Workshop on Mobility Models for Networking Research*, pp. 41–48, 2008.

[SOM 11a] SOMMER C., DRESSLER F., "Using the right two-ray model? A measurement based evaluation of PHY models in VANETs", *Proceedings of ACM MobiCom*, pp. 1–3, 2011.

[SOM 11b] SOMMER C., GERMAN R., DRESSLER F., "Bidirectionally coupled network and road traffic simulation for improved IVC analysis", *IEEE Transactions on Mobile Computing*, vol. 10, no. 1, pp. 3–15, 2011.

[SOM 11c] SOMMER C., ECKHOFF D., GERMAN R. *et al.*, "A computationally inexpensive empirical model of IEEE 802.11 p radio shadowing in urban environments", *Wireless On-Demand Network Systems and Services (WONS), Eighth International Conference*, pp. 84–90, 2011.

[SON 17] SONGCHITRUKSA P., SUNKARI S., UGALDE I. *et al.*, "Interlinking Vissim and ns-3 for Connected-Vehicle Simulation: Case Study of Intelligent Dilemma Zone Avoidance" *Journal of the Transportation Research Board*, (in press) 2017.

[STU 10] STUBING H., BECHLER M., HEUSSNER D. *et al.*, "simTD: A Car-to-X system architecture for field operational tests", *IEEE Communications Magazine*, vol. 48, no. 5, pp. 148–154, 2010.

[WAN 09] WANG S.Y., CHOU C.L., "NCTUns tool for wireless vehicular communication network researches", *Simulation Modelling Practice and Theory*, vol. 17, no. 7, pp. 1211–1226, 2009.

[WIN 07] WINNER consortium, D1.1.2, WINNER II channel models, WINNER European Research project Public Deliverable, 2007.

8

Highway Road Traffic Modeling for ITS Simulation

8.1. Introduction

Future Intelligent Transportation Systems (ITS) will rely heavily on new data transmission technologies, which will transform vehicles into actual communication hubs. Among such ITS-enabling technologies, those realizing direct vehicle-to-vehicle (V2V) communication are the most disruptive. They are expected to interconnect vehicles into self-organized networks whose functions are fully distributed, and provide an important complement to the current mobile communication architecture, which is instead based on a radio access infrastructure that centralizes all data exchanges. Matter-of-factly, upcoming 5G networks will integrate traditional cellular and vehicle-to-vehicle (V2V) direct communication into a unifying framework that will allow users to benefit from the best of the two worlds. Specifically, V2V communication is expected to support services that require rapid, stateless, multicast transmissions, including, for example, collision avoidance, cooperative awareness or localized data dissemination.

After years of research and development, the deployment of V2V communication is now close: standards such as IEEE 802.11-2012[1], IEEE 1609[2], OSI CALM-M5[3] and ETSI ITS-G5[4] have been finalized, and regulators in the USA plan to enforce

Chapter written by Marco GRAMAGLIA, Marco FIORE, Maria CALDERON, Oscar TRULLOLS-CRUCES, Diala NABOULSI.

1 IEEE 802.11-2012 – Wireless LAN Medium Access Control (MAC) and Physical Layer (PHY) Specifications. Note that IEEE 802.11p was integrated into the 2012 version of IEEE 802.11.

2 IEEE 1609 – Family of Standards for Wireless Access in Vehicular Environments (WAVE).

3 OSI Standard 21215 – Communications Access for Land Mobiles (ITS-CALM-M5).

4 ETSI Standard EN 302 665 – Intelligent Transportation Systems (ITS).

V2V radio interfaces on all new vehicles by 2017 [MAS 11]. Extensive field tests are also in progress: representative examples are the German sim^{TD} project in Europe and the Ann Arbor Safety Pilot in Michigan, USA.

However, the cost and complexity of large-scale experiments still make computer simulations the method of choice for the performance evaluation of networking solutions based on V2V communication. Dependable simulations are therefore paramount to a proper evaluation of network protocols and algorithms intended for vehicular environments. In this context, the correct modeling of road traffic has been repeatedly proven to be a crucial aspect [FIO 08, BAI 09, UPP 14]. In addition, as well as being dependable, simulations need to be reproducible: this makes the public availability of road traffic datasets as important as their realism [JOE 12].

In this chapter, we focus on the representation of road traffic for the simulation of highway vehicular networks based on V2V communication technologies. In section 8.2, we review different open-access approaches to highway road traffic modeling for network simulation. In section 8.3, we include in our review an original fine-tuned measurement-based mobility model. In section 8.4, we compare the diverse approaches in terms of the instantaneous vehicular network connectivity they induce in a practical case study, i.e. highway segments in the conurbation of Madrid, Spain. The results shed light on the fact that a fine-tuned measurement-based model yields a level of detail in the mobility representation that is necessary for a reliable simulation under generic network settings. In section 8.5, we then leverage such a model to derive fundamental properties of the highway vehicular network connectivity, which are shown to hold across heterogeneous road traffic scenarios. In section 8.6, we conclude the chapter by discussing the networking implications of our investigation.

8.2. Road traffic models

The recognized impact of road traffic modeling on the simulation of vehicular networks has led to a significant effort in increasing the realism of road traffic traces used by network simulators.

A first approach consists of recording real-world movements of vehicles, typically by logging their position via GPS. These mobility traces can then be replayed in simulation to reproduce the actual road traffic. However, datasets of this type are limited to specific vehicles, e.g. fleets of taxis [HUA 07] or buses [DOE 10]; this clearly limits the scope of the networking studies they can support, both in terms of scale and penetration of the V2V communication technologies. In addition, there is currently no real-world dataset of vehicular mobility that is specific to the highway environment we consider in this chapter.

The generation of synthetic vehicular traces is the *de facto* standard approach to road traffic modeling. Here, special attention has been paid to urban road traffic: in this case, the generation process relies on microscopic road traffic simulators, such as SUMO [KRA 12] or VanetMobiSim [HÄR 11]. These are fed with *(i)* real-world road topologies that describe the layout and features (e.g. direction, number of lanes, speed limit, signalization) of all streets in the considered scenario and *(ii)* origin–destination matrices collected from user surveys [UPP 14, RAN 03] or from roadside detectors [COD 15] that describe the macroscopic flows followed by the vehicles within the urban environment. Several datasets were generated using such an approach, e.g. Zurich [RAN 03], Cologne [UPP 14] or Luxembourg [COD 15].

However, the dynamics of traffic over urban regions are not comparable to those of highways: the former are characterized by vehicles traveling at low or medium speed and often crossing intersections regulated by traffic lights or roundabouts; the latter feature instead high speeds and frequent overtaking. In the case of highway road traffic, three basic components are required for the generation of synthetic vehicular mobility:

– the *highway scenario* is a description of the highway road segment to be simulated; it includes the segment span, number of lanes, and speed limits on each lane, and the presence of inflow or outflow ramps;

– the *traffic input feed* is the characterization of the inflow of vehicles at the beginning of the considered highway segment; it models the inter-arrivals of vehicles on each lane, as well as their initial speed;

– the *mobility model* is the mathematical representation of the driving behavior of vehicles that travel on the simulated segment; the model is typically microscopic, i.e. it determines the acceleration or deceleration of each vehicle separately, based on the surrounding conditions.

The vehicular networking literature is very heterogeneous when it comes to the implementation of the three components mentioned above. Some works propose to use aggregate statistics to describe vehicle inflows, while others employ fine-grained, per-vehicle traffic count data. Some works employ stochastic models of drivers' behavior, whereas others leverage complex microscopic models. Many works neglect the presence of entry and exit ramps, whereas others consider them. Next, we propose a limited set of prototypal models that capture the vast majority of those employed in the literature. Specifically, we focus on the traffic input feed and on the mobility model, since they are independent of a specific context and can be employed across different highway scenarios. We will instead detail the specific highway scenario we consider in our discussion later on in section 8.4.1.

8.2.1. *Traffic input feeds*

All traffic input feeds fall in between two extreme approaches. The first is that of *real* traffic input feeds, and it imposes that vehicles enter the simulated highway segment according to some real-world traffic counts. Such traffic counts shall provide information on the actual transit of each vehicle, and include data such as the lane, the precise (e.g. order of millisecond) timestamp, the speed and possibly the length or type of the vehicle. Such high-precision data is challenging to collect: usually, real-world counts are obtained via induction loops, infrared counters or cameras, which are programmed to provide coarse-grained data. This is because the public transportation authorities that gather such information are generally interested in aggregate measures on, e.g. the number of vehicles transiting on a road, their average speed or the percentage of heavy vehicles, so as to detect major alterations of traffic conditions. Collecting fine-grained real-world counts implies changing the setup of the devices, so that they store data on each transiting vehicle separately.

The second extreme approach is that of *synthetic* traffic input feeds, where probability distributions are used to model the inter-arrival or inter-spacing of subsequent vehicles. Such distributions can be then used to generate the inflow into the simulated highway segment. Many varied distributions have been employed in the literature, which include deterministic [AKH 15, FEL 14], exponential [KHA 08] and lognormal [WIS 07] arrivals, up to generative models for mixture distributions [GRA 14].

Intermediate situations between these two approaches are also possible. Specifically, synthetic traffic input feeds can be trained on real-world traffic counts. In this case, traffic counts are leveraged to infer experimental distributions of the inter-arrival times; then, theoretical distributions are fitted on the experimental ones. Since inter-arrivals are not constant over time (consider, e.g. rush hours and overnight traffic conditions), such a process is repeated over disjointed time windows of duration w [BAI 09, MON 12]. Clearly, the shorter the time window w, the more accurate the input feed but the larger the number of theoretical distributions needed to model the feed.

Drawing from the classification above, we consider a set of five input feeds. In the following, `real` indicates a real traffic input feed, where vehicles are inserted into the simulation using their actual lane, timestamp and speed. By `synthetic-`w, we denote instead four different versions of synthetic input feeds. There, w is the time window over which the traffic count data is aggregated: 5 minutes, 10 minutes, 30 minutes or one hour. The inter-arrival times for the feed `synthetic-`w are exponentially distributed as follows:

$$f_w(t) = \lambda_w e^{-\lambda_w t},$$

where $\lambda_w = \frac{\mathcal{N}_w}{w}$ is the average number of vehicles per unit of time. The starting lanes are randomly selected in the `synthetic-`w case, and vehicles enter them with a uniformly random speed extracted from a distribution:

$$f_w(s) = \mathcal{U}\left(\mathcal{S}_w^{min}, \mathcal{S}_w^{max}\right).$$

Specifically, $\mathcal{S}_w^{min} = 0.9\,\bar{\mathcal{S}_w}$, $\mathcal{S}_w^{max} = 1.1\,\bar{\mathcal{S}_w}$ and $\bar{\mathcal{S}_w}$ is the average inflow speed observed during time window w. For the sake of consistency with common practices in the literature [BAI 09], we train the λ_w and $\mathcal{S}_w$ parameters of `synthetic-`w models from measurement data.

8.2.2. *Mobility models*

The mobility models employed in the literature on highway vehicular simulation are many and varied. They range from simplistic constant-speed representations [YOU 08, BAI 09] to complex dedicated implementations [AKH 15, FEL 14]. We tested the following representative methodologies.

The `unstructured` approach simply assigns a speed to each vehicle entering the simulated highway segment, and allows each vehicle to travel at that constant velocity along the whole segment. The speed is typically extracted from a uniform probability distribution [YOU 08], which may be calibrated using real-world measurements [BAI 09]. The second option, closer to reality, is the one we adopt in our discussion. In any case, this model clearly neglects all interactions among vehicles, and possibly allows them to overlap during movement. It is, however, a computationally inexpensive approach that has been largely adopted in vehicular networking research.

The `SUMO` approach leverages the SUMO tool, i.e. the *de facto* standard open-source software for the simulation of vehicular mobility [KRA 12]. SUMO implements microscopic car-following and lane-changing models. The former is Krauss' model [KRA 97], which regulates each vehicle's acceleration as a function of the distance to the leading one, the current speed, the safety distance or the acceleration and deceleration profiles. The latter is Krajzewicz's model [KRA 09], which allows vehicles to make overtaking and lane-change decisions, considering the position and speed of nearby vehicles on different lanes. These models provide a much more complex and realistic representation of the movement of each vehicle within the traffic flow. An important remark is that Krauss' and Krajzewicz's models are employed with their standard parameterization, as done in virtually all works that rely on SUMO for their simulations.

8.3. Fine-tuned measurement-based model

In addition to the mobility models outlined in section 8.2.2, we also consider an original fine-tuned mobility model that builds on measurement data. The model, first presented in [GRA 16], leverages the IDM [TRE 00] and MOBIL [TRE 02] microscopic representations of the car-following and lane-changing behaviors, respectively. Although widely adopted in the vehicular networking literature, the IDM and MOBIL are invariably used with their default settings. Instead, the mobility model we introduce here performs an accurate tuning of IDM and MOBIL parameters, so as to better mimic real-world driving behaviors on highways.

Model	Parameter	Meaning	Value
IDM	a	Maximum acceleration	1 m/s^2
IDM	b	Maximum (absolute) deceleration	2.5 m/s^2
IDM	v_i^{max}	Maximum desired speed	~ $f_V(v)$
IDM	Δx^{safe}	Minimum distance	1 m
IDM	Δt_i^{safe}	Minimum safe time headway	~ $f_T(\Delta t)$
MOBIL	p	Politeness factor	0.5
MOBIL	a_L	Bias acceleration (left)	0 m/s^2
MOBIL	a_R	Bias acceleration (right)	0.2 m/s^2
MOBIL	k	Hysteresis threshold factor	0.3

Table 8.1. *IDM and MOBIL parameter settings*

Table 8.1 summarizes the calibration adopted by the model. Specifically, the default values indicated in the original works [TRE 00, TRE 02] are found to work well for the acceleration a, deceleration b, politeness factor p and minimum bumper-to-bumper distance Δx^{safe}. The other parameters instead have to be tuned so as to avoid instability in the synthetic road traffic [GRA 16], as detailed below.

Maximum desired speed. Vehicles can be introduced in the simulation at the time and with the speed defined by the real-world traffic count dataset. However, we also need to configure their maximum desired speed v_i^{max}, i.e. the velocity that vehicle i would keep if alone on the highway.

To that end, we recall that speeds measured from real-world traffic in free flow traffic conditions are representative of desired speeds. Indeed, free flow indicates complete lack of road traffic congestion: vehicles in free flow state have very little interaction, and travel at velocities around their maximum desired speed. Free flow speed distributions can thus be extracted for each lane of the target highway scenario: exemplary Probability Density Functions (PDF) are shown in Figure 8.1(a), 1.1(b) and 1.1(c), for the reference highway scenarios introduced in section 8.4.1. The PDFs are separated by lane, as drivers traveling on different lanes tend to have dissimilar maximum desired speeds. Interestingly, the distributions are different across lanes of the same highway, as faster drivers tend to stay on the leftmost lanes[5]. Moreover, all PDFs have Gaussian shapes, with fitted theoretical distributions indicated by solid lines in Figure 8.1.

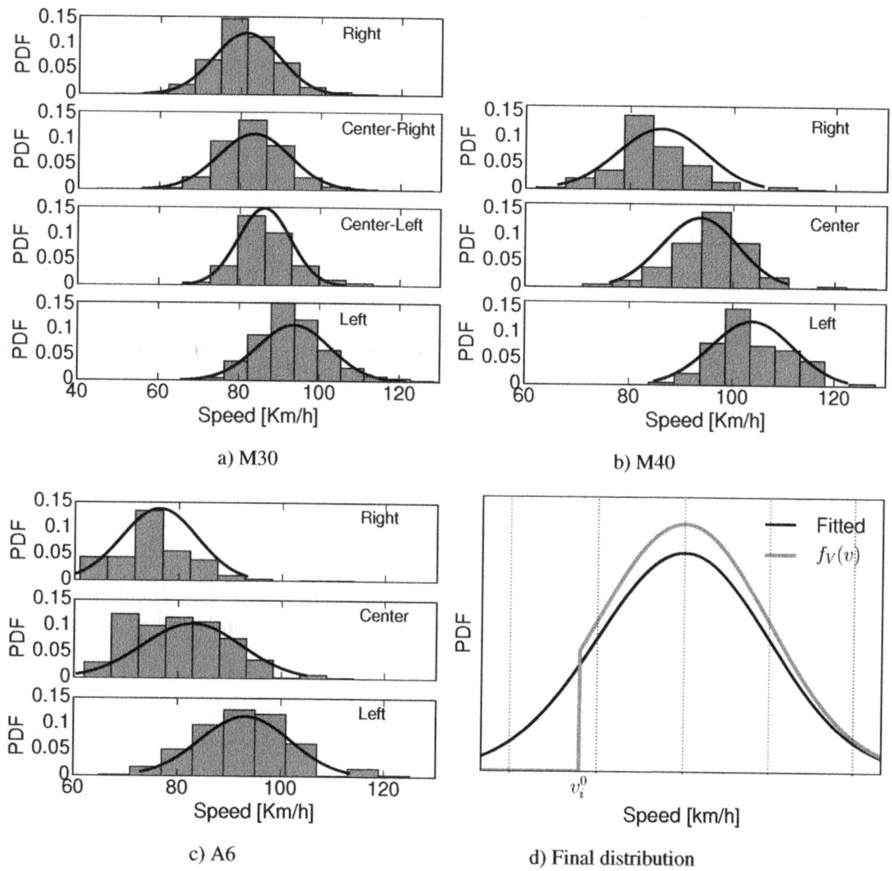

Figure 8.1. *Calculation of the maximum desired speed v_i^{max}. (a,b,c) Empirical and fitted distributions of the free flow speed on each lane of M30, M40 and A6, respectively. (d) Example of per-vehicle truncation and normalization of the fitted distribution, so that only values larger than the initial speed v_i^0 are considered for v_i^{max}, $\forall i$. Figure from [GRA 16]. For a color version of this figure, see www.iste.co.uk/hilt/transportation.zip*

5 The reference scenarios are located in Spain, where rightmost lane occupancy rules are enforced and overtaking occurs on the left. See section 8.4.1 for details.

The PDFs mentioned above allow us to model the maximum desired speeds as Gaussian-distributed random variables, whose mean $\mu_{h,l}$ and standard deviation $\sigma_{h,l}$ vary depending on the highway h and lane l considered. In fact, this is not sufficient: as drivers traveling on a same lane are not all equal, we adapt the final v_i^{max} distribution on a per-vehicle basis as follows:

$$f_V(v) = \begin{cases} 0, & v < v_i^0 \\ \frac{\sqrt{2}\, exp(-(v-\mu_{h,l})^2/2\sigma_{h,l}^2)}{\sigma_{h,l}\sqrt{\pi}\left[1+erf((v_i^0-\mu_{h,l})/\sigma_{h,l}\sqrt{2})\right]}, & v \geq v_i^0. \end{cases} \quad [8.1]$$

The expression in [8.1] truncates the Gaussian distribution at the speed v_i^0 recorded in the real-world traffic count data for vehicle i and renormalizes it. Figure 8.1(d) provides a graphical example. The initial velocity of i, i.e. v_i^0, becomes a lower bound to v_i^{max}: this ensures that the maximum desired speed of a vehicle i is never lower than its initial v_i^0, which would conflict with the real-world measurements.

Minimum safe time. The minimum safe time headway Δt_i^{safe} is known to vary across real-world scenarios, in the range from 0.9 s [NHT 01] to 3 s [WHI 14]. In the proposed mobility model, we infer its value, on a per-vehicle basis, from road traffic measurements.

Specifically, the inter-arrival times between vehicles recorded in real-world traffic can be directly related to the Δt_i^{safe} values. However, this only holds when the road traffic is very dense, and inter-vehicle spacing actually maps to safety distances. More formally, according to traffic flow theory, the traffic density ρ on lane l of highway h can be expressed as follows:

$$\rho_{h,l} = \frac{1}{L + \Delta t_{h,l}^{safe} v_{h,l}}, \quad [8.2]$$

where L is the average length of the vehicles, $v_{h,l}$ is the average speed and $\Delta t_{h,l}^{safe}$ is the average safe time headway [CHO 14]. From density $\rho_{h,l}$, we can compute the vehicular flow $q_{h,l} = \rho_{h,l} \cdot v_{h,l}$, which results in:

$$\Delta t_{h,l}^{safe} = \frac{1}{q_{h,l}} - \frac{L}{v_{h,l}}. \quad [8.3]$$

Expression [8.3] directly relates $\Delta t_{h,l}^{safe}$ to the maximum value of the flow $q_{h,l}$ and average speed $v_{h,l}$. The maximum flow $q_{h,l}$ can be inferred from a real-world traffic count dataset by identifying the time interval during which a speed breakdown occurs on all lanes. The average speed $v_{h,l}$ is extracted from the same data as the average velocity of vehicles in free flow conditions, and L is the average vehicle length.

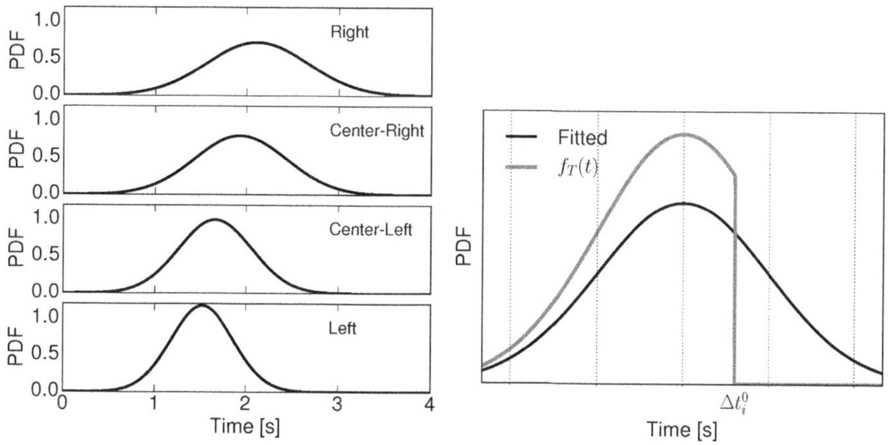

a) M30 b) Final distribution

Figure 8.2. *Calculation of the minimum safe time headway Δt_i^{safe}. (a) Reference distributions of the typical safe time headway on each lane of M30, as inferred by the experimental flow, speed and inter-arrival information contained in the traffic count dataset. (b) Example of per-vehicle truncation and normalization of the reference distribution, so that only values smaller than the initial inter-arrival time Δt_i^0 are considered for Δt_i^{safe}, $\forall i$. Figure from [GRA 16]*

The reference Gaussian distribution of safe time headway is then assigned the computed mean $\Delta t_{h,l}^{safe}$. The standard deviation $\sigma_{h,l}$ can be set such that the minimum inter-arrival time recorded in the real-world traffic count dataset represents the 0.99 quantile of the distribution, i.e. three standard deviations. An example of the resulting per-lane distributions is provided in Figure 8.2(a) for one of the reference highway scenarios detailed in section 8.4.1: we remark that the values of $\Delta t_{h,l}^{safe}$ obtained for all lanes (2.11, 1.93, 1.66 and 1.52 s from the rightmost to the leftmost lane, respectively) are well aligned with those found in the literature [TRE 00, WHI 14, NHT 01].

As a final step, similar to what done for the maximum desired speed, a per-vehicle distribution is to be determined from the lane-dependent reference ones. In this case, the final Δt_i^{safe} distribution is given by:

$$f_T(\Delta t) = \begin{cases} \dfrac{\sqrt{2}\, exp(-(\Delta t - \Delta t_{h,l}^{safe})^2/2\sigma_{h,l}^2)}{\sigma_{h,l}\sqrt{\pi}\left[1+erf((\Delta t_i^0 - \Delta t_{h,l}^{safe})/\sigma_{h,l}\sqrt{2})\right]}, & t \leq \Delta t_i^0 \\ 0, & t > \Delta t_i^0, \end{cases} \quad [8.4]$$

where Δt_i^0 is the initial inter-arrival time of vehicle i recorded in the traffic count dataset. As shown in Figure 8.2(b), [8.4] allows Δt_i^0 to become the upper bound to Δt_i^{safe}. This ensures that no vehicle enters the simulation with an inter-arrival time that is lower than its minimum safe time headway.

Lane change bias and hysteresis threshold. In our highway scenarios, the default MOBIL settings result in a traffic that is highly skewed towards the left lane, which thus suffers from unrealistic congestion. We ran a comprehensive campaign to identify the combination of right (a_R) and left (a_L) lane change bias, and lane change hysteresis threshold factor (k) that grants quasi-stationary traffic over the different lanes. Such consistent ingress and egress per-lane properties were obtained for a_R = 0.2 m/s^2, a_L = 0 m/s^2 and k = 0.3. Interestingly, the lane change bias favors movements to the right in absence of a clear preference among lanes, which is in compliance with road regulation in Spain.

The mobility model arising from all the fine-tuning above is indicated as `IDM` in the following. A software implementation and sample datasets of synthetic highway road traffic generated with this model are open to the research community[6].

8.4. Comparative analysis of road traffic models

In this section, we provide a comparative evaluation of the different strategies for synthetic highway traffic generation presented before. We thus test combinations of `real` and `synthetic-`w traffic input feeds with `unstructured`, `SUMO` and `IDM` mobility models. More precisely, we consider a reference highway scenario, detailed in section 8.4.1, and study the effect of the diverse approaches on the connectivity of the vehicular network built on V2V communication, according to the metrics presented in section 8.4.2. The results of this approach are summarized in section 8.4.3.

8.4.1. *Case study scenarios*

The highway scenario considered in our comparative evaluation is that of highways around the conurbation of Madrid, Spain. Fine-grained real-world traffic counts were collected by the Madrid City Council on M30, M40 and A6 for the purpose of our study. The data describes individual vehicle transits (including vehicle speed and type) with a 100 ms time accuracy, and covers heterogeneous traffic conditions from very sparse overnight traffic to rush hour congestion.

The different traffic input feed and mobility models presented in sections 8.2 and 8.3 are fed with this real-world measurement data. The `real` feed matches the data, whereas in the `unstructured` feed the initial speed is derived from a probability distribution fitted on the data. In `IDM` mobility model, the target speed and minimum gap between subsequent vehicles are calculated as described in section 8.3.

6 Available at `http://www.it.uc3m.es/madrid-traces`

In addition to the highway settings, a reliable study of vehicular networks also requires a proper representation of the RF signal propagation model. Indeed, such a model determines whether vehicles are capable of communicating via V2V technologies. We thus extract V2V communication distance from a state-of-the-art propagation model [ABB 15], considering the transmission power is set to 20 dBm, a received signal strength threshold of −91 dBm and a reliability of .99. Shadowing effects due to nearby vehicles are considered as well, via an additional path loss when the latter obstruct the line-of-sight.

8.4.2. *Connectivity metrics*

Our investigation is based on a protocol-independent approach that focuses on instantaneous connectivity metrics of vehicular networks. The metrics describe the global structure of the vehicular network and measure its level of connectivity or fragmentation. They are formalized as follows.

At each time instant t, we represent the network as an undirected graph $G(\mathbb{V}(t), \mathbb{E}(t))$. Each vertex in the set $\mathbb{V}(t) = \{v_i(t)\}$ maps to the vehicles i in the network at time t, and each edge in the set $\mathbb{E}(t) = \{e_{ij}(t)\}$ connects $v_i(t)$ and $v_j(t)$ if a V2V communication link exists between vehicles i and j at time t. We also denote as $\mathcal{N}(t) = \|\mathbb{V}(t)\|$ the number of vertices in the graph, i.e. the *number of vehicles* in the scenario, at time t.

Let us define a component $C_m(t) = G(\mathbb{V}_m(t), \mathbb{E}_m(t))$ as a subgraph of $G(\mathbb{V}(t), \mathbb{E}(t))$, where $\mathbb{V}_m(t) \subset \mathbb{V}(t)$ includes all and only the vertices corresponding to vehicles that can reach each other via direct or multi-hop communication at time t. Equivalently, $\mathbb{E}_m(t) = \{e_{ij}(t) \mid v_i(t), v_j(t) \in \mathbb{V}_m(t)\} \subseteq \mathbb{E}(t)$. We denote as $\mathcal{S}_m(t) = \|\mathbb{V}_m(t)\|$ the size of the component $C_m(t)$. Since components are disjointed by definition, $\mathcal{C}(t) = \|\{C_m(t)\}\|$ is the *number of components* appearing in the network at time t. The number and size of components in the network at each time instant will be our network connectivity metrics.

The *component availability* and *component stability* metrics study large connected components emerging in the network, which are especially interesting as they allow for significant multi-hop communication opportunities. In particular, the two metrics focus on (i) the presence and (ii) the temporal fluctuations of such large components. Formally, we refer to the largest component appearing in the network at time t as $C_{max}(t) = G(\mathbb{V}_{max}(t), \mathbb{E}_{max}(t)) = C_m(t) \mid m = \arg_n \max \mathcal{S}_n(t)$. Then, $\mathcal{S}_{max}(t) = \|\mathbb{V}_{max}(t)\|$ is the size of the largest component at the same time instant. The normalized value of $\frac{\mathcal{S}_{max}(t)}{\mathcal{N}(t)}$ at each instant will be our reference metric for the study of the component availability, whereas its temporal variations will be leveraged to analyze the component stability. More precisely, the component stability is assessed through the correlograms of $\mathcal{S}_{max}(t)$: correlograms are derived by dividing

$\mathcal{S}_{max}(t)$ time series into time windows, and computing the temporal autocorrelation at different lags, for each window.

In the remainder of the chapter, we will drop the time notation for the sake of brevity and refer all metrics to a generic time instant. We will thus use $\mathcal{N}$ to indicate the number of vertices in the network, $\mathcal{C}$ for the number of components and $\mathcal{S}_{max}$ the largest component size.

8.4.3. *Results*

We first assess the impact of mobility modeling on the global network connectivity, expressed as the component availability, i.e. the ratio between $\mathcal{S}_{max}$ and $\mathcal{N}$. Figure 8.3 portrays smoothed scatter plots that refer to different combinations of traffic input feed and mobility models. All plots show the metrics as functions of the road traffic density, in vehicles per km. We highlight remarkable differences across plots, as follows.

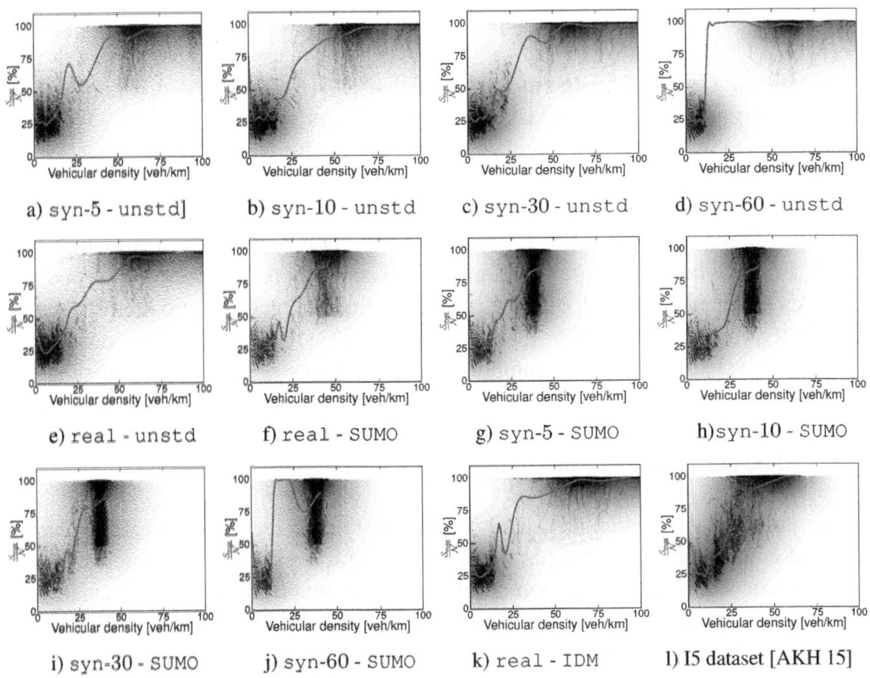

a) syn-5 - unstd]
b) syn-10 - unstd
c) syn-30 - unstd
d) syn-60 - unstd
e) real - unstd
f) real - SUMO
g) syn-5 - SUMO
h) syn-10 - SUMO
i) syn-30 - SUMO
j) syn-60 - SUMO
k) real - IDM
l) I5 dataset [AKH 15]

Figure 8.3. *The relative availability $\mathcal{S}_{max}/\mathcal{N}$ versus the vehicular density $\mathcal{N}$. The red line denotes the average. Figure from [GRA 15]. For a color version of this figure, see www.iste.co.uk/hilt/transportation.zip*

First, the parameter w (in minutes) strongly influences the connectivity and availability metrics. While Figure 8.3(a), 8.3(b) and 8.3(c) show a comparable and realistic behavior; using synthetic traffic with $w = 60$, in Figure 8.3(d), yields an abrupt transition between the disconnected (~20% percent availability) and fully connected (~100% percent availability) phases. Equivalent considerations hold when synthetic traffic is combined with microscopic mobility, see, for example, the striking difference between Figure 8.3(g) and 8.3(j). We conclude that *an exceedingly coarse inflow granularity risks completely losing the state transitions that occur in real-world traffic, as well as the associated connectivity and availability states.* Unfortunately, w is often a non-configurable parameter decided by the data providers, who are typically only interested in rough aggregates of the inflow traffic for statistical purposes.

Second, the use of `SUMO` appears to cause issues with the observed metrics. All plots where `SUMO` is used to model the vehicular mobility show that the mobility generator is just unable to insert all the vehicles in the simulation. This is clear when looking at Figure 8.3(f) and 8.3(g)–8.3(j). While `unstructured` and `IDM` attain a peak traffic density of about 70 vehicles per km, `SUMO` never exceeds 40 vehicles per km. This is a parameterization issue: the default settings of Krauss' model do not allow accommodating high inflows observed in the real world, which forces `SUMO` to delay the insertion of a vehicle until Krauss' model safety requirements are fulfilled. In turn, this affects network connectivity and availability.

These results prove that *using a validated microscopic model of vehicular mobility is not sufficient to obtain a realistic representation of road traffic: the parameterization of the model is extremely important, and a careless setting can lead to biased simulation outcomes.* Clearly, this is not a problem of Krauss' model *per se*. In order to prove it, we also show the connectivity and availability metrics obtained using the mobility dataset described in [AKH 15], which was generated using `SUMO` with customized (but undisclosed) parameterization. Figure 8.3(l) shows similar trends to those obtained with `unstructured` and `IDM`.

Third, an interesting observation is that *a very simple constant-speed simulator using synthetic (but sufficiently detailed) traffic input feed results in a network connectivity and availability comparable to those attained by much more complex models.* Figure 8.3(a), 8.3(e) and 8.3(k) shows precisely this effect.

Fourth, we stress that the highway road traffic dataset in [AKH 15] describes traffic in a different scenario, i.e. Interstate highway 5 (I5) in CA, USA. Still, the connectivity and availability scatter plots and mean curves are identical to those of our reference scenarios in Spain. This result allows us to speculate on the general validity of our findings, which could apply to different highway environments.

The correlograms of S_{max} in Figure 8.4 display the temporal variation of the largest connected component in the network: they map to the component stability metric. Here, we only display a subset of the results, for the sake of brevity and since

w did not appear to influence the component stability. Again, SUMO, in Figure 8.4(b) and 8.4(d), shows a very different trend due to the maximum density issue we already discussed. However, the important result here is that the unstructured mobility model exhibits clear limitations. Figure 8.4(a) and 8.4(c) proves how the lack of interaction among vehicles in these models results in correlograms that differ from that obtained with IDM, in Figure 8.4(e). In the latter model, drivers are forced to adjust their speed according to the surrounding road traffic conditions, which leads to well-known phenomena, such as synchronized traffic: in turn, the global reduction of speed and queuing of vehicles noticeably improve connected component lifetime. We conclude that *a simplistic representation of microscopic mobility does not impact network-wide metrics, but leads to connected components that may be significantly less stable in time than what would occur in the real world.*

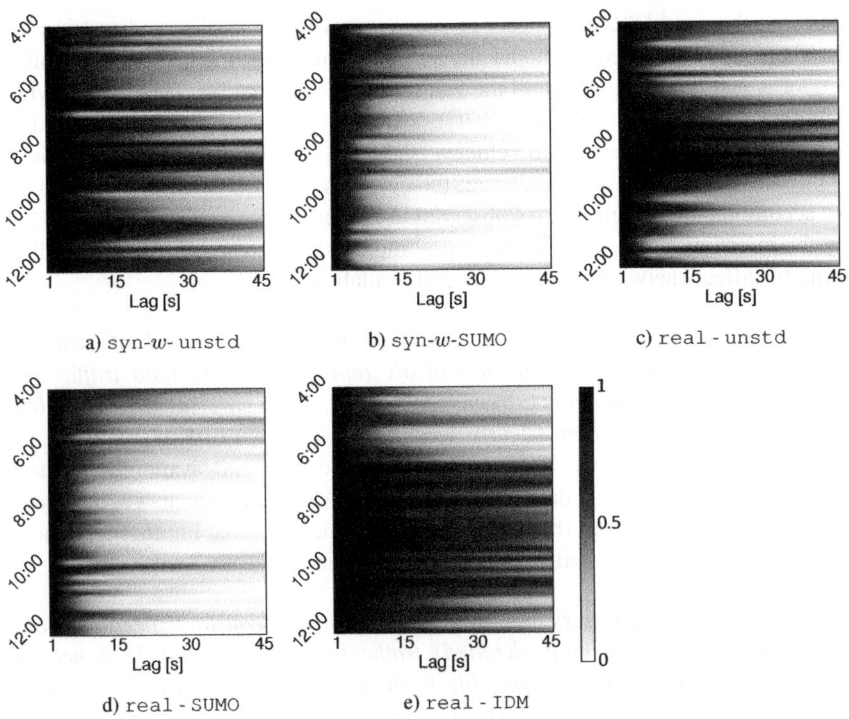

Figure 8.4. *$S_{max}/\mathcal{N}$ correlograms. Figure from [GRA 15]*

8.5. Fundamental properties of highway vehicular networks

In this section, we leverage the most realistic highway traffic representation among those evaluated in section 8.4, i.e. the IDM mobility model tuned on a real data feed to derive key properties of vehicular connectivity in highway environments.

Specifically, we investigate the existence of general laws explaining the fluctuations of vehicular network connectivity as a function of two system parameters: the V2V communication range, denoted as R, and the road traffic density $\mathcal{N}$.

Figure 8.5 portrays the evolution of $\mathcal{C}$ and $\mathcal{S}_{max}$ versus $\mathcal{N}$. Each plot refers to a different R and shows the average behavior recorded in the M30 highway scenario (black solid line), as well as the dispersion around that mean (0.05–0.95 quantile range, as a light gray region). The vertical dashed lines roughly separate $\mathcal{N}$ ranges corresponding to sparse overnight traffic, typical free flow traffic and synchronized congested traffic.

The dynamics of both $\mathcal{C}$ and $\mathcal{S}_{max}$ are related to $\mathcal{N}$. The largest component size, in the bottom plots, features a clear positive correlation with $\mathcal{N}$. The number of components, in the top plots, instead displays a skewed bell shape. A clear *three-phase connectivity in* $\mathcal{N}$ emerges, under any R, from Figure 8.5. The three phases, or behavioral regions, are as follows:

I) For low $\mathcal{N}$, $\mathcal{S}_{max} \sim 1$ and $\mathcal{C}$ grow linearly with $\mathcal{N}$: the network is very sparse and increasing the number of vehicles just means adding more isolated nodes, i.e. singleton components;

II) Once a first critical $\mathcal{N}$ threshold is reached (denoted by the leftmost red dotted vertical line "A" in the plots), a second behavior ensues. Namely, $\mathcal{S}_{max}$ grows super-linearly with $\mathcal{N}$ and $\mathcal{C}$ decreases sub-linearly with $\mathcal{N}$. The reason is that, beyond a critical vehicular density, new cars tend to join existing components or even bridge them into larger ones;

III) The third region is attained after a second $\mathcal{N}$ threshold (the rightmost red dotted vertical line "B" in the plots) is passed. There, $\mathcal{S}_{max} \sim \mathcal{N}$ and $\mathcal{C} \sim 1$, i.e. the vehicular network is fully connected into a single component whose size matches the number of vehicles on the highway segment.

The behavior mentioned above is invariant over different values of the communication range R. Yet, the value of R greatly affects the critical $\mathcal{N}$ thresholds that trigger phase changes, which are anticipated for larger values of R.

The fact that the M30 curves show a moderate 0.05–0.95 quantile interval around the mean allows us to theorize that considering one single road traffic parameter, i.e. $\mathcal{N}$, is enough to properly characterize the vehicular connectivity in all situations encountered during a typical working day. An interesting corollary to this observation is that other features, such as the daytime, day of the week, number of lanes, speed limits or presence of ramps are only responsible for minor variability in the connectivity. Such an observation also holds for the actual road traffic conditions (i.e. free flow to synchronized or jammed traffic, which are known to induce, for example, major speed variations), which are not decisive to connectivity region transitions.

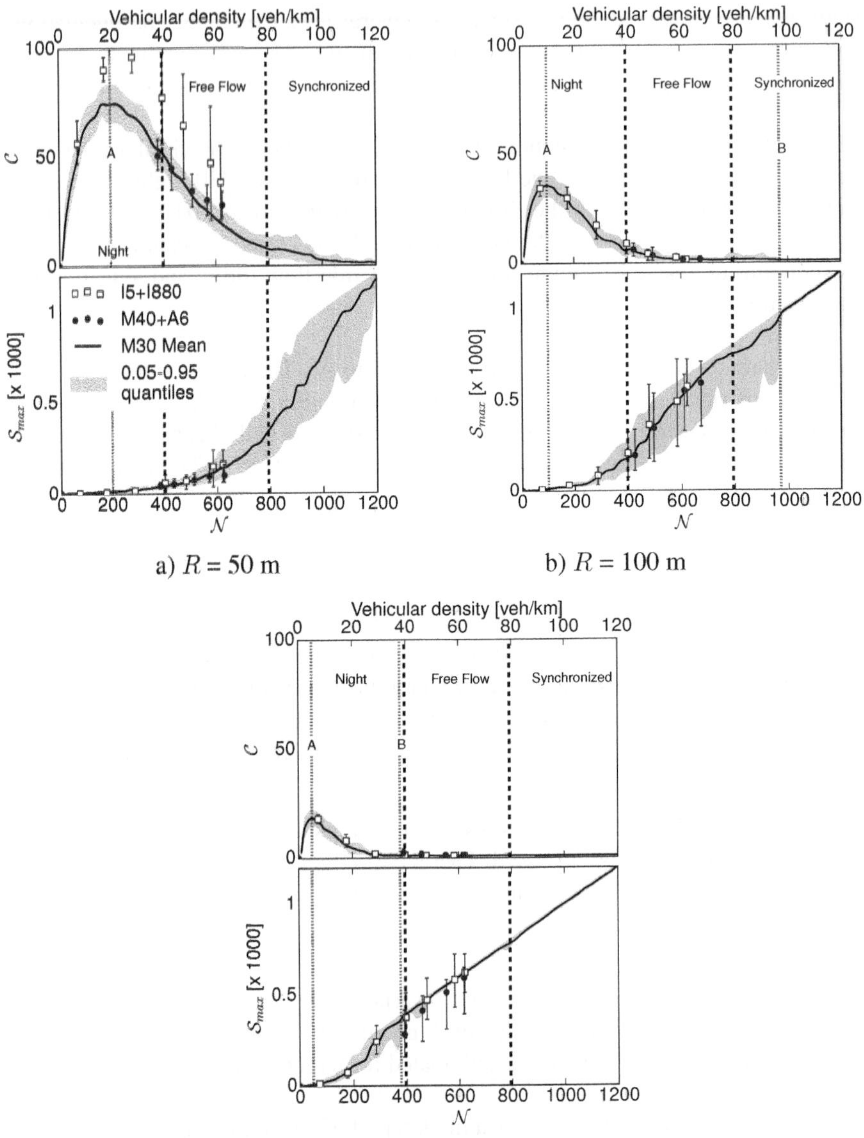

Figure 8.5. *C and S_{max} versus the number of nodes $\mathcal{N}$ for the M30, M40, A6, I5 and I880 datasets, for different R values. For a color version of this figure, see www.iste.co.uk/hilt/transportation.zip*

Figure 8.5 also includes the $\mathcal{C}$ and $\mathcal{S}_{max}$ recorded in the M40 and A6 highway scenarios (represented as filled circles in the graphics), as well as in additional I5 and I880 highway scenarios (empty squares). The latter correspond to the highway environments considered in [AKH 15], where measurement data from the US Freeway Performance Measurement System (PeMS) is fed to a properly calibrated SUMO simulator to generate synthetic road traffic. For the M40, A6, I5 and I880 scenarios, dots represent the mean $\mathcal{C}$ and $\mathcal{S}_{max}$ values and error bars denote 0.05 and 0.95 quantiles. We remark that the majority of M40, A6, I5 and I880 dots fall very close[7] to the mean behavior observed in the M30 case, and their 0.05–0.95 quantile ranges tend to correspond to those of M30. Therefore, *we conclude that the same three-phase connectivity dynamics in $\mathcal{N}$ hold for all of the highway scenarios we consider.* Moreover, the impact of R on the network connectivity is equivalent in all such scenarios. Once more, these observations allow us to speculate that the three-phase connectivity law may have general validity for vehicular networks in highway environments.

8.6. Discussion and conclusions

The results presented in section 8.4 demonstrate that a specialized highway mobility model like `IDM`, fine-tuned on a `real` data feed, is *necessary* for a faithful representation of road traffic in network simulations. If such a requirement is not met, significant errors emerge in the V2V communication-based connectivity, which can then propagate to the performance of network solutions.

Surprisingly, even a state-of-the-art mobility simulator such as `SUMO` cannot be used straight away, due to an inappropriate parameterization of its mobility model default settings. Instead, an `unstructured` simulator where vehicles travel at constant speed may be sufficient, but only for simulating network solutions that only rely on the availability of large connected components (e.g. best-effort data dissemination or collection); when more precise dynamics of the vehicular network must be properly modeled in simulation (e.g. for cooperative awareness or collision avoidance) such an approach can bias the results. We also observe that `synthetic` data can be used to feed simulators, if not aggregated over too large temporal windows w that lose state transitions in real-world traffic.

7 Some difference appears in Figure 8.5(a) for I5 and I880, in terms of the number of components $\mathcal{C}$, when R = 50 m. The reason is the presence of in- and out-flow ramps in these road traffic scenarios: ramps create traffic perturbations that tend to break apart the vehicular network components. However: (i) the impact of ramps on $\mathcal{C}$ in that plot is not dramatic, and results are only slightly shifted from those of M30, M40 and A6; (ii) in- and out-ramps do not appear to affect the size of large components, as seen for $\mathcal{S}_{max}$ (see Figure 8.5(a) bottom plot); (iii) the effect of ramps on $\mathcal{C}$ disappears when R grows (in Figure 8.5(b) and 8.5(c)).

The following discussion in section 8.5 allows us to conclude that the topology of highway vehicular networks is driven by two major factors, i.e. the V2V communication range and the road traffic density. More precisely, such an interdependence occurs through an invariant three-phase relationship that connects connectivity and road traffic density (not to be confused with the road traffic state).

Overall, these results shed light on the fundamental dynamics of vehicular network topologies, and have clear implications in the design and performance evaluation of adaptive networking solutions intended for vehicular environments.

8.7. Bibliography

[ABB 15] ABBAS T., SJÖBERG K., KAREDAL J. *et al.*, "A measurement based shadow fading model for vehicle-to-vehicle network simulations", *International Journal of Antennas and Propagation*, Article ID 190607, 2015.

[AKH 15] AKHTAR N., ERGEN S.C., OZKASAP O., "Vehicle mobility and communication channel models for realistic and efficient highway VANET simulation", *IEEE Transactions on Vehicular Technology*, vol. 64, no. 1, pp. 248–262, 2015.

[BAI 09] BAI F., KRISHNAMACHARI B., "Spatio-temporal variations of vehicle traffic in VANETs", *Proceedings of the Sixth ACM International Workshop on VehiculAr InterNETworking – VANET '09*, pp. 43–52, 2009.

[CHO 14] CHO S., CRUZ R., RAO R. *et al.*, "Time-gap based traffic model for vehicular traffic flow", *2014 IEEE 79th Vehicular Technology Conference (VTC Spring)*, pp. 1–5, May 2014.

[COD 15] CODECA L., FRANK R., ENGEL T., "Luxembourg SUMO Traffic (LuST) Scenario: 24 hours of mobility for vehicular networking research", *Vehicular Networking Conference (VNC), 2015 IEEE*, pp. 1–8, December 2015.

[DOE 10] DOERING M., PÖGEL T., PÖTTNER W.-B. *et al.*, "A new mobility trace for realistic large-scale simulation of bus-based DTNs", *Proceedings of the 5th ACM Workshop on Challenged Networks – CHANTS '10*, pp. 71–74, 2010.

[FEL 14] FELICE M.D., BAIOCCHI A., CUOMO F. *et al.*, "Traffic monitoring and incident detection through VANETs", *2014 11th Annual Conference on Wireless On-demand Network Systems and Services (WONS)*, pp. 122–129, April 2014.

[FIO 08] FIORE M., HÄRRI J., "The networking shape of vehicular mobility", *Proceedings of the 9th ACM International Symposium on Mobile Ad Hoc Networking and Computing – MobiHoc '08*, pp. 261–272, 2008.

[GRA 14] GRAMAGLIA M., FIORE M., CALDERON M., "Measurement-based modeling of interarrivals for the simulation of highway vehicular networks", *IEEE Communications Letters*, vol. 18, no. 12, pp. 2181–2184, 2014.

[GRA 15] GRAMAGLIA M., FIORE M., "On the level of detail of synthetic highway traffic necessary to vehicular networking studies", *Vehicular Networking Conference (VNC), 2015 IEEE*, pp. 17–24, December 2015.

[GRA 16] GRAMAGLIA M., TRULLOLS-CRUCES O., NABOULSI D. *et al.*, "Mobility and connectivity in highway vehicular networks: a case study in Madrid", *Computer Communications*, vol. 78, pp. 28–44, 2016.

[HÄR 11] HÄRRI J., FIORE M., FILALI F. *et al.*, "Vehicular mobility simulation with VanetMobiSim", *Simulation*, vol. 87, no. 4, pp. 275–300, 2011.

[HUA 07] HUANG H.-Y., LUO P.-E., LI M. *et al.*, "Performance evaluation of SUVnet with real-time traffic data", *IEEE Transactions on Vehicular Technology*, vol. 56, no. 6, pp. 3381–3396, 2007.

[JOE 12] JOERER S., SOMMER C., DRESSLER F., "Toward reproducibility and comparability of IVC simulation studies: a literature survey", *IEEE Communications Magazine*, vol. 50, no. 10, pp. 82–88, 2012.

[KHA 08] KHABAZIAN M., ALI M.K.M., "A performance modeling of connectivity in vehicular ad hoc networks", *IEEE Transactions on Vehicular Technology*, vol. 57, no. 4, pp. 2440–2450, 2008.

[KRA 97] KRAUSS S., WAGNER P., GAWRON C., "Metastable states in a microscopic model of traffic flow", *Physical Review E*, vol. 55, pp. 5597–5602, 1997.

[KRA 09] KRAJZEWICZ D., "Kombination von taktischen und strategischen Einflüssen in einer mikroskopischen Verkehrsflusssimulation", *Fahrermodellierung in Wissenschaft und Wirtschaft*, pp. 104–115, 2009.

[KRA 12] KRAJZEWICZ D., ERDMANN J., BEHRISCH M. *et al.*, "Recent development and applications of SUMO – simulation of urban mobility", *International Journal on Advances in Systems and Measurements*, vol. 5, no. 3, 2012.

[MAS 14] MASON J., LAWDER D., "Obama backs highway fund fix, touts 'talking' cars", *The New York Times*, Reuters, 2014, Accessed 6 September 2014.

[MON 12] MONTEIRO R., SARGENTO S., VIRIYASITAVAT W. *et al.*, "Improving VANET protocols via network science", *2012 IEEE Vehicular Networking Conference (VNC)*, pp. 17–24, November 2012.

[NHT 01] NHTSA, "Distance behaviour on motorways with regard to active safety – a comparison between adaptive-cruise-control (ACC) and driver", *ESV*, 2001.

[RAN 03] RANEY B., CETIN N., VÖLLMY A. *et al.*, "An agent-based microsimulation model of Swiss travel: first results", *Networks and Spatial Economics*, vol. 3, no. 1, pp. 23–41, 2003.

[TRE 00] TREIBER M., HENNECKE A., HELBING D., "Congested traffic states in empirical observations and microscopic simulations", *Physical Review E*, vol. 62, pp. 1805–1824, 2000.

[TRE 02] TREIBER M., HELBING D., "Realistische Mikrosimulation von Strassenverkehr mit einem einfachen Modell", *Arbeitsgemeinschaft Simulation (ASIM)*, Rostock, September 2002.

[UPP 14] UPPOOR S., TRULLOLS-CRUCES O., FIORE M. *et al.*, "Generation and analysis of a large-scale urban vehicular mobility dataset", *IEEE Transactions on Mobile Computing*, vol. 13, no. 5, pp. 1061–1075, 2014.

[WHI 14] WHITE J., "2014 Rules of the road", *CyberDrive Illinois*, 2014.

[WIS 07] WISITPONGPHAN N., BAI F., MUDALIGE P. *et al.*, "Routing in sparse vehicular ad hoc wireless networks", *IEEE Journal on Selected Areas in Communications*, vol. 25, no. 8, pp. 1538–1556, 2007.

[YOU 08] YOUSEFI S., ALTMAN E., EL-AZOUZI R. *et al.*, "Analytical model for connectivity in vehicular ad hoc networks", *IEEE Transactions on Vehicular Technology*, vol. 57, no. 6, pp. 3341–3356, 2008.

9

F-ETX: A Metric Designed for Vehicular Networks

9.1. Introduction

Due to their inherent characteristics, including self-organization, scalability, mobility and the fast changing transmission channel quality, vehicular ad hoc networks (VANET) address specific challenges. Vehicles move on the road network according to traffic patterns and they do not rely on a limited battery capacity. Vehicle-to-Vehicle (V2V) communications rely on the cooperation with each other to build opportunistic wireless networks. Since vehicles move with a wide range of speeds according to traffic patterns, the network topology is characterized by a potentially high dynamic. The road environment (e.g. urban, suburban and motorway) also plays a key role in the disturbance of the transmission channel.

Distributed applications require the cooperation of nodes, but they are bounded by connectivity and reliability issues. Those are partially solved by routing protocols, which ensure an end-to-end communication with a multi-hop relaying technique. To this end, protocols compute and share local information on the direct neighborhood to determine the best end-to-end path. A relevant challenge for routing protocols is the selection of the best kind of estimator to obtain reliable information on local links. Indeed, routing performance in terms of end-to-end delay and packet delivery ratio depends on the reliability of the selected path. The traditional hop count metric relates the cost of a path to the number of hops required to reach a destination. However, De Couto *et al.* [DEC 03b] have demonstrated the inefficiency of such a technique in wireless networks.

Chapter written by Sébastien BINDEL, Benoit HILT and Serge CHAUMETTE.

Link Quality Estimators (LQE) have been developed in order to fix the intrinsic limitations related to the hop count metric. They take into account either the signal quality or the lossy and the dynamic of a link to assess its quality. As discussed, in [BAC 12], LQEs have to meet four requirements, including: (i) the energy consumption, (ii) the accuracy, (iii) the reactivity and (iv) the stability of the estimation. However, for an LQE to be suitable for vehicular networks, additional requirements must be considered. The first one deals with the support of the node mobility since a vehicle may change its speed according to the speed limits and the traffic patterns. Regarding the signal disturbance, environment impacts the transmission channel quality. Finally, Zamalloa and Krishnamachari [ZAM 07] showed that the quality of a link can be split into three regions, connected, transitional and disconnected. In the connected region, a link has a high probability of having a high packet reception rate. In the disconnected region, a link has a high probability of having a low packet reception rate. The transitional region is an intermediate region characterized by an unstable link quality. The main challenge for an LQEs is accurately assessing the link quality regardless of the current region.

In order to build a thorough sample and assess the link quality, current estimators maintain an estimation window that stores received packets. Carpa *et al.* [CER 05a] invested the related challenges in order to assess the Packet Reception Ratio (PRR). They determine that a window can have a small size if the PRR is high or low, but a larger size is needed in other cases. However, current estimators keep a fixed estimation window size regardless of the PRR and cannot provide a reliable assessment if the link quality is situated in the transitional region. As a result, current LQEs have limited effectiveness in vehicular networks. To address the problem of the link assessment in vehicular networks, the Fast Expected Transmission Count (F-ETX) estimator has been developed [BIN 15b]. Unlike current estimators, it uses a dynamic window size fitting according to the packet loss occurrences. From experiments, we have observed that such an estimation provides only a snapshot of the quality. This remains insufficient since the quality trend is not taken into account to compare links among each other. Figure 9.1 shows the quality of a couple of links. If a link selection mechanism relies on a short-term estimation, it will continuously switch the best one among available links during the 80th and the 90th second. In contrast, a long-term estimation can highlight a tendency of the link quality while one grows (Link #2) and another decreases (link #1). We argue that a multi-estimator approach as suggested in [BAC 10b] [REN 11] is a better approach. As a result, F-ETX has been extended with three additional estimators assessing distinct features of a link in order to assess the link quality and determine the link state [BIN 15a]. We have also developed a framework in order to integrate the F-ETX within routing protocols. This chapter provides a detailed investigation of the metric. We provide a theoretical analysis of the window estimation and show its relationship with the quality assessment. Then, we describe the design of the F-ETX including window management algorithms and multi-estimators, and outline the framework to integrate the metric into a routing protocol. Finally, we prove its usefulness through realistic simulations.

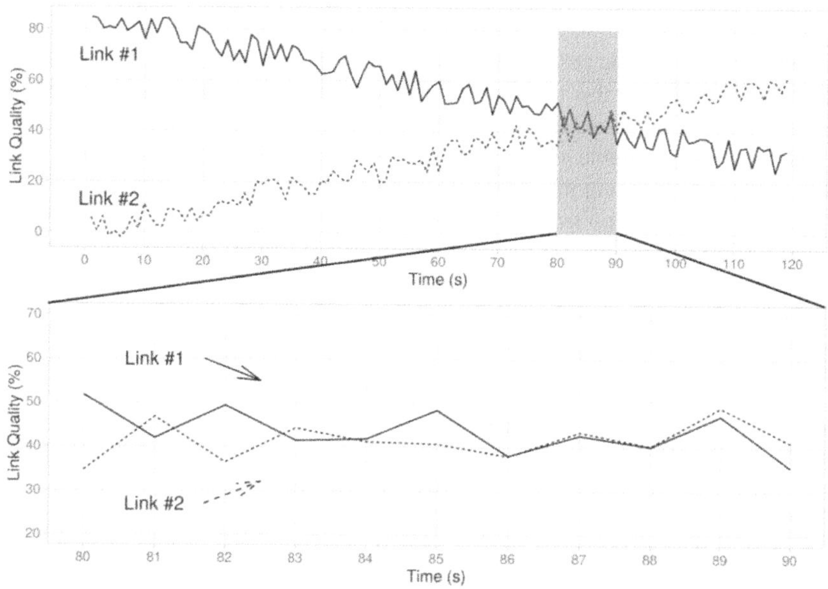

Figure 9.1. *The link selection problem*

The remainder of this chapter is organized as follows. In section 9.2, the literature related to the LQEs is described. In section 9.3, we perform an analysis of the estimation by regarding its impact on the reactivity and the accuracy of the estimator. In section 9.4, we detail the F-ETX metric, including the couple of algorithms managing the window size and each estimator, and we outline the framework to integrate the metric into a routing protocol. In section 9.5, we describe simulation settings and depict results in section 9.6.

9.2. Link quality estimators

Link monitoring and measurement are a fundamental building block to assess the quality of wireless links. This information is used by several algorithms, e.g. for routing decision and group formation to build a substantial knowledge of the neighborhood. The literature related to this topic has been well investigated and several estimators have been developed. In this section, we summarize the literature and highlight the main challenges. LQEs can be classified into two categories: (i) hardware-based and (ii) software-based.

9.2.1. *Hardware-based LQE*

Hardware-based estimators profit from measures made at the physical layer by the hardware to assess the link quality. The quality is assessed as soon as a packet is received, without specific cost since the measurement is performed by the hardware. Such estimators assess the link quality by performing measurements on received packets. An estimation from a hardware-based estimator that correlates with the PRR is considered as a suitable metric. Two types of estimators can be distinguished, classical metrics exploiting the signal properties and a novel generation of estimators which retrieve information from the decoding process of the Direct-Sequence Spread Spectrum (DSSS) and the Orthogonal Frequency-Division Multiplexing (OFDM) techniques.

9.2.1.1. *Signal property-based*

The first one is the Received Signal Strength Indicator (RSSI) based on the measurement of the power reception of the incoming frame. Practical experiences have shown that RSSI can provide an accurate estimation when a link has good quality [SRI 06]. Srinivasan et al. [SRI 06] showed that above a RSSI value ($-87dBm$), the PRR is consistently high (99%). In [SRI 10], the same authors observed that the standard deviation of PRR is weak over a short time span. A variation of the RSSI can change a good link to a bad link if it operates near to the noise floor. A simple reading of the RSSI value is not sufficient to determine the related PRR, since they are not sufficiently correlated. The second one is the Signal-to-Noise Ratio (SNR). For a given modulation scheme, the bit error rate can be computed with the SNR, which can be extrapolated to the packet error rate and so the PRR [ZAM 07]. Unlike the RSSI, which is the sum of the power of the signal and the noise, the SNR indicator determines how strong the signal is compared to the ambient noise. As a result, SNR is a better candidate than the RSSI indicator. However, experiments have shown that correlation with the PRR is not deductible at all, and Lal *et al.* [LAI 03] depreciate its use when the link has an intermediate link quality. The Link Quality Indicator (LQI) was introduced in the IEEE 802.15.4 [IEE 16] for low rate wireless networks. Experimental works led by Lui et Cerpa [LIU 14] show that the assessment provided by the LQI has the best matching compared to the RSSI and the SNR. However, in the transitional region a simple reading of the LQI is insufficient to determine the PRR since its variance is too important. Boano *et al.* [BOA 09] suggest using the variance to distinguish good and bad links.

9.2.1.2. *Decoding event-based*

Heinzer *et al.* [HEI 12] developed a metric dealing with DSSS decoding process to measure the Chip Error Per Symbol (CEPS). However, the correlation between the assessment given by this metric and the PRR can be approximated by a linear fitting. To overcome this drawback, a novel metric called BLITZ was designed, also dealing with the DSSS decoding process [SPU 13]. Unlike CEPS, which performs on the payload, BLITZ relies on a measurement on the frame preamble used to synchronize the sender and the receiver. Experimental results show the better performance of BLITZ compared to the other metrics, but the experimentation environment was limited to a simple transmitter and a receiver in the same collision domain. Gabteni *et al.* [GAB 14] developed a link state indicator that analyzes decoding errors of the OFDM reception process. Called Link State Forwarding Indicator (LSFI), it is able to predict future link disruptions.

9.2.2. *Software-based*

Software-based estimators retrieve information from uppers layers, e.g. MAC and Net, to determine whether an expected packet will be received or not. These estimators are usually classified in three categories: (i) PRR-based, (ii) RNP-based and (iii) Score-based.

9.2.2.1. *PRR-based*

This type of estimator is based on successive PRR measurements to determine the link quality. Classical approaches maintain a window to monitor and sample traffic over the link. Cerpa *et al.* [CER 05a] advise the maintenance of a narrow window if a link has a low or high PRR. On the other hand, the links with a medium PRR must be monitored with a larger window to enhance the estimation accuracy. Woo and Culler [WOO 03] designed the WMEWMA technique to smooth the PRR estimation. This technique is based on the EWMA filter that applies an exponential weight to give more importance to the newest or the oldest data. These estimators share the same drawbacks by assessing the link quality through the traffic of the downlink. Indeed, they cannot take into account losses of the uplink; this is why RNP-based estimators were suggested.

9.2.2.2. *RNP-based*

RNP-based estimators monitor both the downlink and uplink to assess the link quality. Cerpa *et al.* suggested the RNP (Required Number of Packet transmissions) estimator counting the average number of retransmissions required to deliver a packet [CER 05b]. The protocol requires the use of an ARQ (Automatic Repeat reQuest) technique for counting the number of failed and succeed transmissions. The Expected Transmission Count (ETX) metric was designed by De Couto *et al.* [DEC 03a] and takes into account the delivery ratio (computed from the average number of transmitted packets successfully received) and the reverse delivery ratio (computed from the average number of successfully received ACKs) to assess the link quality. Unlike RNP, which uses data traffic (passive method), ETX has to monitor the link with an active monitoring technique.

9.2.2.3. *Score-based*

Score-based estimators combine multi-estimators in order to assess the link quality and determine the link state. Baccour *et al.* [BAC 10] designed a hybrid metric called F-LQE, based on a multi-estimator approach, each assessing the packet delivery ratio, the link asymmetry level, the link stability and the channel quality. These estimators are aggregated into a single metric following a fuzzy logic method. In addition, they implemented F-LQE into the Collection Tree Protocol (CTP) routing protocol and proved its effectiveness in wireless sensor networks [BAC 15]. The Holistic Packet Statistic (HoPS) metric suggested by Renner *et al.* [REN 11] incorporates four estimators, namely short term, long term, absolute deviation and trend estimation. However, an intrinsic problem of the use of this filter limits the agility of estimators. It also has the disadvantage of requiring a large amount of traffic to train the estimators and consequently increases the detection time of link state changes.

9.2.3. *Discussion*

Classical hardware-based estimators measure the signal quality to determine the reception state of the upcoming packet. Experiments have proven the inability of such metrics to provide a fine grain of link quality, since the correlation with the PRR is not deductible. Computed from successfully received packets, these estimators may overestimate the link quality by not considering lost packets. A novel type of estimator extracts and analyzes information retrieved from the decoding process. Even if they are more accurate than classical approaches, they require information retrieved from specific radio chips.

On the other hand, software-based estimators assess the link quality according to the application point of view, i.e. the successful packet reception ratio or packet transmitted. Unlike hardware-based estimators software-based estimators, especially RNP-based, are able to assess both parts of a link (uplink and downlink) to ensure a more reliable assessment. Experimental works have confirmed this observation. As a result, such estimators have been well used by routing protocols. Beside, score-based metrics provide a multi-faced assessment to obtain a reliable link quality. This state-of-the-art is summarized in Table 9.1.

9.3. Analysis of legacy estimation techniques

In this section, we address the issue concerning estimation windows of RNP-based estimators by focusing our attention on the fulfillment process and computation techniques. This lays the foundation for us to understand and analyze performances of current assessment techniques. We regard only active traffic monitoring where nodes monitor the links of their neighbors by broadcasting probe packets.

Since RNP-based estimators assess both sides of a link, the quality assessment relies on two information sources. Indeed, ETX-like estimators compute two ratios: (i) d_f counting the number of packets successfully received by a neighbor and (ii) d_r counting the number of received packets from a neighbor. Figure 9.2 shows the link monitoring scheme of ETX-like estimators. Several techniques have been retained to count received packets. Two kinds must be considered and are the main purpose of the next section.

Type[1]	Categories	Name	Technique	Location	Link[2]
H	Signal properties	RSSI	Signal Strength	Receiver	←
H	Signal properties	SNR	Signal to Noise ratio	Receiver	←
H	Signal properties	LQI	Error between the ideal constellation and the received signal	Receiver	←
H	Decoding event	CEPS, BLITZ	DSSS decoding process	Receiver	←
H	Decoding event	LSFI	OFDM decoding process	Receiver	←
S	PRR-based	PRR	Average	Receiver	←
S	RNP-based	RNP	Average	Sender	↔
S	RNP-based	ETX	Average	Receiver	↔
S	Score-based	F-LQE	Fuzzy logic	Receiver	←
S	Score-based	HoPS	Heuristic	Receiver	←

Table 9.1. *LQE review*

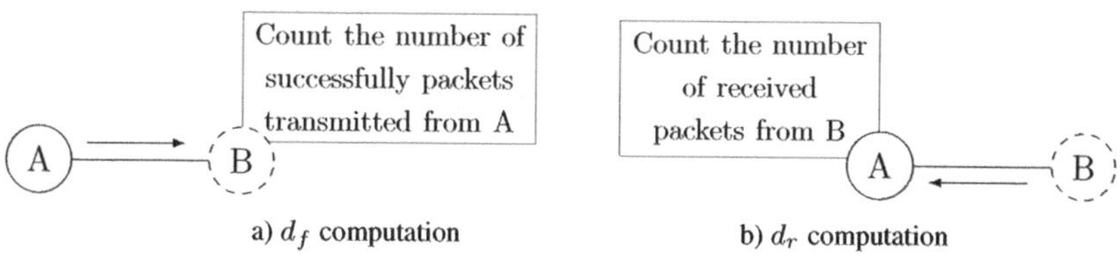

Figure 9.2. *Link monitoring scheme*

9.3.1. *Type of window*

In order to monitor the link and to perform measures, estimators use a window mechanism. This forms a representative sample of the transmitted and received traffic. According to [DEC 03a] and [QUA 11], samples can be built from temporal or sequential information.

9.3.1.1. *Temporal*

Temporal information was introduced by De Couto *et al.* [DEC 03a]. The d_f ratio is computed from the number of probe packets received by a neighbor. To this end, nodes must periodically exchange the number of packets received from the neighbor. The computation of the d_f ratio is detailed in the following equation:

$$r(t) = \frac{Count(t - w, t)}{w/t}. \qquad [9.1]$$

$Count(t - w, t)$ is a function counting probe packets received during a period w and w/t is the number of probe packets which should be received. The freshness of the d_f ratio relies on the fixed period w determining the sample size. The main drawback of this approach is the exchange of the d_f ratio. If a probe packet is lost, a receiver cannot determine whether its probe packet has been successfully transmitted. As a result, it assumes the transmitted packet as lost, even if it has been successfully received. Since the d_f ratio is exchanged periodically through probe packets sent, nodes cannot exchange their current values. Indeed, nodes send the d_f ratio corresponding to the last exchange and not the current one.

9.3.1.2. *Sequential*

Rather, sequential information provides an affordable solution to determine ratios. To this end, a sequence number is used as an ID and is assigned exclusively to probe packets. A novel and fresh approach has been developed in [QUA 11], which is actually implemented in the Better Approach To Mobile Ad-hoc Networking (BATMAN) routing protocol. The proposed approach avoids the exchange of the d_f ratio by changing the retransmission policy, where only the emitter node is responsible for the computation of ratios. This is computed from the assumption that $d_f \times d_r$ represents the ratio that a transmission is successfully received and acknowledged. According to Figure 9.3, each probe packet is retransmitted by the receiver in order for the originator of the probe packet to compute the delivery ratio.

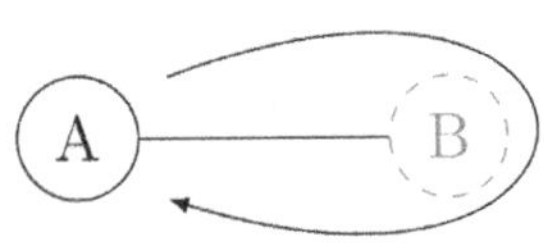

Figure 9.3. *Novel assessment of the d_f ratio*

With this approach, nodes are able to assess the current link quality, since information about the two ratios is acquired within the current period. On the other hand, the number of transmission increases, because each probe packet must be forwarded. Avoiding an infinity retransmission is ensured with the retransmission policy. Probe packets have to contain three specific fields: the node's address that creates the probe packet (AddrOrign), the address of the last forwarder (AddrPrev) and the sequence number (SN). The retransmission policy is described in algorithm 1.

Algorithm 1 Retransmission policy

```
INPUT: packet: received packet
INPUT: node_addr : receiver's address
  if packet.AddrOrig = packet.AddrPrev then
      Computedr()
      packet.AddrPrev ← node_addr
      SendPacket(packet)
  else if packet.AddrOrig = node_addr then
      Computedf()
      DropPacket(packet)
  else
      DropPacket(packet)
  end if
```

9.3.2. *Window analysis*

In this section, we address the problem of the window size since it impacts both the convergence time and the accuracy of the estimator. We focus our attention on sequential windows, fulfilled periodically with probe packets. The filling of the window depends on its size and the sending period. However, reducing the sending period has a negative impact on the network performance, since it reduces the bandwidth allocated to data. To this end, the only way to change the convergence time of a window is to adapt its size.

The size impacts the time to fulfill the window, so it determines the time to declare a link with a maximum quality or detect a disruption. In this section, we investigate the computational techniques that rely on a window mechanism. We focus our attention on the assessment technique used by ETX and show the benefits and the limits of such a method. The quality assessed by ETX takes into account both the d_f and d_r ratios and is computed as $\frac{1}{d_f \times d_r}$. Figure 9.4 shows the impact of the mean filter on the convergence time to declare a link with a maximal quality. A larger size implies a longer time to declare a link with a maximal quality, since it increases the time to fulfill the window. Concerning the stacked density function, we observe that an estimator with a larger window size is able to assess the link quality with more values. However, the distribution function is more situated on the left, this means that more values to describe low qualities (< 50) are obtained.

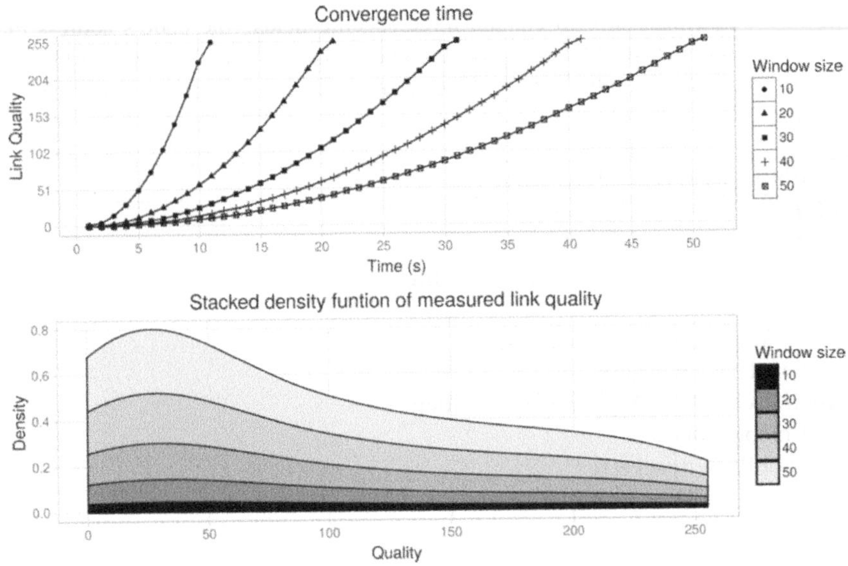

Figure 9.4. *ETX: link emergence study*

The window size also determines the number of observations required to assess a link as being at its maximum quality. It also impacts the estimation accuracy, since it determines the observation number of the sample. We consider each attempt transmission as a Bernoulli trial. Thus, the reception state of a packet can be described as a binary value, meaning its successful reception or its loss. Therefore, the sample can be described as a binary word whose length is the sample size. Determining the estimation accuracy can be done with combinatorial analysis. Since ETX uses the mean filter to compute the two ratios, the ranking is not considered and the sample can be seen as a combination with allowed repetitions. Let n describe the binary reception state and k the sample size, the total number of combinations is described by:

$$C_k^{n+k-1} = \frac{(n+k-1)!}{k!(n-1)!}, \qquad [9.2]$$

$$C_k^{k+1} = \frac{(k+1)!}{k!} = k+1\cdot$$

With a larger window size, ETX is able to assess the link quality with more values than a small window size. However, as shown in Figure 9.4, this increases the convergence time of the estimator. We have also observed, on the density function plot, the inequality distribution of estimated values, with most situated on the left (lowest values). As a result, ETX is not able to be both reactive and accurate, because it uses a fixed window size.

9.4. The F-ETX metric

Several efforts have been made to develop trustworthy link quality estimators. Most of them have been developed and tested in Wireless Sensor Networks (WSN)s. In section 9.3, we have observed that for an estimator a static window size implies a trade-off between accuracy and reactivity. Their effectiveness is limited in mobile environments, since they do not deal with the short span of link lifetimes. Besides, mobile nodes can evolve in different environments with specific mobility patterns, which leads to unpredictably disturb the radio channel.

A novel metric called F-ETX has been proposed to deal with the problem of the link quality assessment in mobile networks. The metric is composed of four estimators, each assessing a specific feature of the link and allows a multi-faced assessment. The metric is able to assess the link quality and determine the link state in order to prevent future events, such as a link disruption. F-ETX avoids the problems related to the use of static window size for traffic monitoring by using a dynamic management of the size. We regard the packet loss as a relevant event to reduce or extend window size. To this end, the metric owns two algorithms to manage the window size, each one assigned to a specific job: size reduction and size growth.

9.4.1. *Window management algorithms*

A relevant challenge for LQEs is to provide a quick and an accurate assessment in an unknown and dynamic environment. As depicted in the last section, current solutions imply a trade-off. F-ETX tackles all suggested estimators from all of the state-of-the-art by bringing a dynamic management of the window size. Its main insight is to automatically adapt the accuracy and the reactivity of the estimator. Information concerning the packet losses has been retained as the most relevant to achieve a window size fitting. To this end, F-ETX implements two tight algorithms. The first one is dedicated to the reduction of the window size, in order to improve the estimator reactivity. The second one is able to extend the window size, in order to increase the assessment accuracy.

9.4.1.1. *Window size reducing algorithm*

With the reduction of the window size, the algorithm is able to increase the reactivity of the estimator but also decrease the assessment accuracy. One of the most important features is to detect a link disruption as soon as possible. Let a packet $p \in P$ be a finite whole of observed packets such as $P = \{p_0, p_1, p_2, \cdots, p_{n-1}\}$, with n

being the number of observation. Each observed packet p is labeled according to its reception state, such as a label $L \in [0,1]$, where $L \leftarrow 0$ indicates a loss and $L \leftarrow 1$ a reception. The number of packet considered as received, a, and lost, $\bar{a}$, are computed as follows:

$$n = a + \bar{a},$$

$$a = \sum_{i=1}^{n} L_i, \qquad [9.3]$$

$$\bar{a} = n - \sum_{i=1}^{n} L_i. \qquad [9.4]$$

Thus, the window size n is reduced according to the number of packet considered lost:

$$n = \frac{n}{2^{\bar{a}}}. \qquad [9.5]$$

The key idea is to increase the reduction process, in accordance with the packet loss. If packet losses are sporadic, n is slightly reduced, otherwise, n is significantly reduced. Implementing this algorithm requires some extra considerations. To support traffic monitoring through a window mechanism, packets must contain a sequence number in order to be identified. Since F-ETX is based on an active monitoring, the sending period is used to determine whether a packet can be declared as received or lost. Figure 9.5 shows a study case, where the last expected packet is lost (Sequence Number $N\#8$).

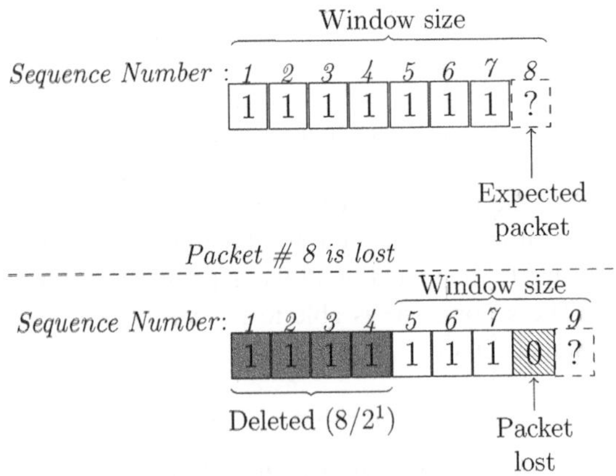

Figure 9.5. *Illustration of the window size reduction algorithm*

9.4.1.2. *Window size growing algorithm*

The following algorithm is able to extend the window size in order to enhance the assessment accuracy. It is triggered after a packet loss and as soon as a novel packet is received. The algorithm proceeds in two steps. The first one is the recovery phase, in which the algorithm extends the window size for each new received packet. The goal is to recover the initial window size before the packet loss. The second one is the link stability sensing, in which the algorithm gropes for the extension of the window. The algorithm tries to extend the window according to the link stability in order to provide a more accurate assessment. The switch between the recovery phase and the link stability sensing phase is triggered by the threshold T_h. Its value is set to the window size before the first packet loss. Until the window size is lower than T_h, the widow size is incremented by one for each new received packet. Indeed, the algorithm tries to recover the last window size before the disruption. Then, in the recovery phase, the window (W) is increased or shifted (left) according to a dedicated counter C, as described in algorithm 2.

Algorithm 2 Window size growing during the recovery phase

```
if C ⩾ W_n/2 then
    W ← W increased by 1
    C ← 0
else
    W ← W Slid by 1
    C ← C + 1
end if
```

The reduction algorithm attempts to detect the disruptions at the earliest by reducing the size of the tight couple of the windows. When the link gets back, the second algorithm tries to recover the initial window size before the last disruption. Upon reaching their initial size, the algorithm gropes for the increase of the window size. During this stage, the size is progressively increased until reaching the maximum window size.

9.4.2. *Multi-assessment approach*

The link quality assessment aims to find the highest throughput link. Renner *et al.* [REN 11] have pointed out the problem of such a method for comparing links (see Figure 9.1). Rather, Renner *et al.* and Baccour *et al.* suggested a metric to be capable of assessing multi-features of a link. These approaches have been developed for WSNs and do not have the required ability to be deployed in mobile networks. Even if previous algorithms enhance the reactivity and the accuracy of the link quality assessment, this is not sufficient. That is why F-ETX implements four estimators, two dedicated to the link quality assessment and another dedicated to determining the link state.

9.4.2.1. *Link quality*

The expected probability that a message is successfully received and acquitted is $d_f \times d_r$. If we consider a packet transmission as a Bernoulli trial (success or fail), the link quality (χ^{LQ}) estimation is determined as follows:

$$\chi^{LQ} = \frac{1}{(1 - d_f) \times (1 - d_r)}. \qquad [9.6]$$

9.4.2.2. *Link quality trend*

This indicator tracks the course of the link quality by computing the variation between the current χ_t^{LQ} and the previous estimation χ_{t-1}^{LQ}. To provide a long-term estimation, this result is averaged with an EWMA filter:

$$\Delta_t^{LQ} = \chi_t^{LQ} - \chi_{t-1}^{LQ}, \qquad [9.7]$$

$$\chi_t^{Trend} = \beta \times \Delta_t^{LQ} + (1 - \beta) \times \chi_{t-1}^{Trend},$$

the coefficient β influences the sensitivity of the estimator. Choosing a small β value is advisable to achieve a long-term estimation. Note that two successive nulls χ^{LQ} indicate a disruption and reset the link quality trend estimator.

9.4.2.3. *Link stability estimation*

We observed that a fine analysis of the window content provides link stability information. Let a binary state $[0, 1]$ representing the reception state of an excepted packet in a window. We denote W_{max} as the maximum window size, W_n the current window size and W_i the ith element in the window. The windows maintained to compute the d_f and d_r probabilities are respectively denoted as W^{d_f} and W^{d_r}. The link stability indicator is computed with an EWMA filter, taking into account the absolute Ξ and the relative stability ξ:

$$\Xi = \frac{\sum_{i=1}^{W_n^{d_f}} W_i^{d_f} + \sum_{i=1}^{W_n^{d_r}} W_i^{d_r}}{2W_{max}},$$

$$\xi = \frac{\sum_{i=1}^{W_n^{d_f}} W_i^{d_f} + \sum_{i=1}^{W_n^{d_r}} W_i^{d_r}}{W_n^{d_f} + W_n^{d_r}}, \qquad [9.8]$$

$$\chi_t^{Stab} = \Xi_t \times \gamma + (1 - \gamma) \times \xi_t.$$

The absolute estimation (Ξ) computed from the maximum window size (fixed value) represents the absolute level of stability of the link. The relative estimation (ξ) computed from the current window size (dynamic value) represents the relative stability. This third estimation gives the level of the link stability according to the current window size. This information is useful, since, for the same absolute value, the relative link estimation gives an additional assessment taking into account losses which occurred recently. Both absolute and relative information are suitable for assessing the link stability. They must be taken into account in the same way. Hence, we advise a γ value fixed at 0.5.

9.4.2.4. *Unidirectional link level*

This last estimator deals with the detection of bidirectional links becoming unidirectional. Current approaches like F-LQE with the ASL estimator track the difference between the uplink and downlink reception rates. Such a method becomes inefficient if the link has a short life time or experiences a high level of packet losses. In this case, windows are not sufficient trained to give a trustworthy estimation. Our method overcomes this limitation by measuring the variation of the up and downlink reception ratios. This makes it independent of the window size and does not require any training period. Let W be a window and W_n^t its size at time t. The variation of the reception ratio provided by the window W at time t is denoted as Δ_t^{Win}. The indicator is given by:

$$\Delta_t^{Win} = \sum_{i=1}^{W_n^t} W_i - \sum_{i=1}^{W_n^{t-1}} W_i,$$

$$\chi_t^{ULL} = \chi_{t-1}^{ULL} \times \lambda + (1-\lambda) \times \varphi(\Delta_t^{d_f}, \Delta_t^{d_r}), \qquad [9.9]$$

$$with\ \varphi(x,y) = \begin{cases} -1 & x < 0 \wedge y > 0 \\ 1 & x > 0 \wedge y < 0 \\ 0 & else \end{cases} \cdot$$

To give a tendency, we advise a λ value fixed at a high value. A link may become unidirectional (e.g. nodes with different transmit power level) if the assessment becomes negative.

9.4.3. **Routing integration framework**

In this section, we describe the framework designed to integrate all estimators into a routing protocol. Each estimator assesses a specific property of a link in order to provide information on its quality and its state. It is a key concept for addressing issues of routing protocols. Current metrics using multi-estimators such as F-LQE [BAC 10] and HoPS [REN 11] compute a scored quality link estimation in order to provide a single value. Even if Baccour *et al.* [BAC 15] have implemented F-LQE in the CTP routing protocol, there are no silver bullets to compute an ultimate single estimation including all assessment provided by estimators.

To solve this issue, we propose a framework integrating each estimation into the routing process. Indeed, each estimator is related to the routing table in order to indicate the link quality and inform us about the link state event occurrence. Based on an active monitoring, each estimation is computed after the reception of a probe packet. Then, they assign their assessment to the associated entry into the routing table. The proposed framework is illustrated in Figure 9.6, including the routing protocol and our metric.

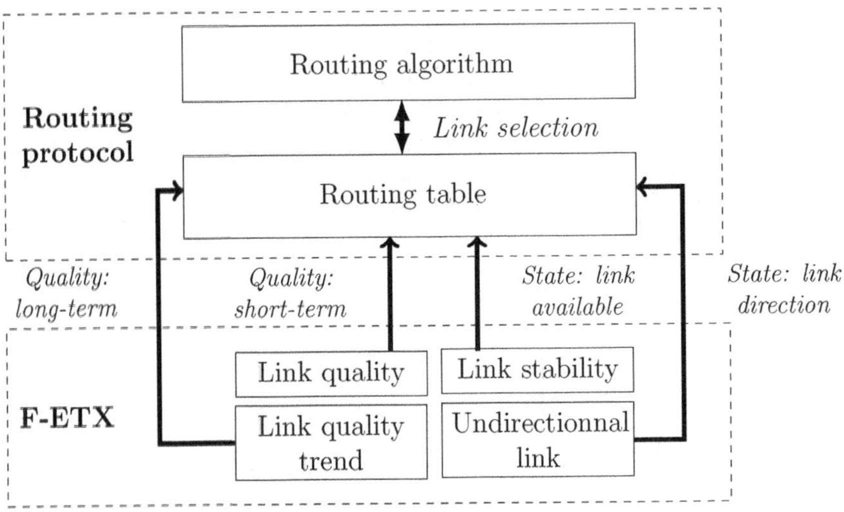

Figure 9.6. *Routing framework*

Since we consider the routing table as the intermediary part between the metric and the outing protocol, each piece of information is stored in the associated entry in the routing table. Then, the routing algorithm selects the best link according to its quality and its state. Thus, we need to define how the quality is assessed and which state is stored in the routing table. In addition, we detail how a routing algorithm may interpret information stored in the routing table.

9.4.3.1. *Local link evaluation*

As pointed out, F-ETX includes four estimators, namely a short-term link quality, long-term link quality, stability estimator and unidirectional link indicator.

As shown in Figure 9.1, it becomes a major concern when two links are close in quality but have opposite trends. We propose a novel link quality assessment merging both the short and the long term:

$$\text{Link quality} = \chi^{LQ} + \chi^{Trend}. \quad [9.10]$$

Taking into account both the short and the long-term estimation makes it possible to penalize the short-term estimation according to its current trend. If a couple of links have a close link quality, the metric is able to select the best one according to its quality trend.

The stability indicator determines if a link is fitted to support data transmission. This information is used to declare a routing entry as enabled or disabled. Indeed, a null estimation brands a routing entry as disable even if the quality is not null. Thus, our estimator also indicates two possible states of the link stability:

– $\chi^{Stab} = 0$: disable routing entry;

– $\chi^{Stab} > 0$: enable routing entry.

The unidirectional link indicator detects transient loses which turn a bidirectional link into an unidirectional link. Besides, persistent unidirectional links can be detected with a direct observation of the d_r ratio indicating the number of packets retransmitted by a neighbor. Thus, our indicator is able to detect the unidirectional property of the link:

– U_p: persistent unidirectional link;

– U_t: transient unidirectional link.

9.4.3.2. *Routing in praxis*

When a packet is received, the routing algorithm is responsible for routing the packet to the destination by selecting the best link. This selection process is performed by selecting the corresponding entry in the routing table. That is why the protocol ranks, for each destination node, the best potential neighbor according to the link quality ($\chi^{LQ} + \chi^{Trend}$). Then, the protocol inspects the stability indicator to determine whether a route is declared available or disable. In the disable case, the algorithm selects the next route and restarts the same approach. At the end, the protocol checks if the link is unidirectional. In the case of a transient state, the route is selected, else the routing algorithm looks for another route.

9.5. Simulation settings

We investigate the performance of the F-ETX metric into two rounds. In the first one, we lead to a performance evaluation between the F-ETX and two current multi-estimators, F-LQE and HoPS. In the second one, we observe the impact of the F-ETX on the routing performance, if it is used as the principal metric. In order to lead our investigation, we define two scenarios, including a realistic mobility pattern and a real signal propagation environment, and simulated with ns-3 [RIL 10].

9.5.1. *First scenario*

The first scenario is used in the performance evaluation of the F-ETX, F-LQE and HoPS. In this scenario, 40 vehicles move in an urban area of $500m \times 500m$ with a Manhattan mobility model 4×4. We set the mean speed of vehicles at $30km/h$ in order to simulate a high speed traffic. From [BEN 12], we fix the channel propagation parameters, with a Three Log Distance Loss Model as a shadowing model and Rayleigh's model as a fast-fading model, to reproduce a realistic urban channel propagation environment. Table 9.2 details propagation environment carefully.

PHY parameters	
Tx/Rx power (dbm) Gain of antenna (dB) Power Detection Threshold (dbm)	0 0 –96
MAC parameters	
Standard Mode Rate adaptation	802.11g OFDM 6 Mpbs ARF
Propagation Loss Parameters	
ThreeLogDistance Exponent 0 Exponent 1 Exponent 2 Distance 0 Distance 1 Distance 2	 2.5 5 10 1 75 114
Nakagami-m Rayleight	 m = 1

Table 9.2. *Signal propagation parameter*

9.5.2. *Second scenario*

The second scenario is used in order to observe the impact of the F-ETX on the routing performance. In this scenario, 50 vehicles move in an urban area of 1 km^2 with a Manhattan mobility model 4×4. We set the mean speed of vehicles at 50 km/h, and limit the minimal speed at 30 km/h. Like in the first scenario, channel propagation parameters are detailed in Table 9.2, which details propagation environment carefully.

9.6. Simulation results

We now describe the results obtained by our experiments. As mentioned previously, we test our approach in a realistic urban environment. We explore the performance of each estimator of the F-ETX metric and its impact on the routing performance.

9.6.1. *Performance of the multi-estimators*

We investigate the performance and the robustness of all estimators of F-ETX. To this end, we compare the performance of our estimators and current metrics namely F-LQE [BAC 10] and HoPS [REN 11]. We set the parameters of each estimator according to [BAC 10] for F-LQE and [REN 11] for HoPS. We fix the parameters of F-LQE as follows. We set the coefficient of the WMEWMA filter used to compute the packet delivery ratio at 0.6. The assessment of the link quality and the link stability requires a history of PRR that we set at 30. A minimal history is maintained at 5 PRR until it reaches 30. For HoPS, we set the parameters as follows. Coefficients are respectively set to 0.9 and 0.997 and their short- and long-term estimations are initialized at 50% for new links. Finally, for F-ETX, we determine the parameters of EWMA filters by setting companion estimators called λ, β and γ to 0.9, 0.1 and 0.5 respectively.

Our main goal is to assess both the agility of estimators by observing their ability to track fluctuation and their accuracy and their robustness. We achieve both temporal and statistical evaluations. Through the temporal experiments, we observe the behavior of estimators in order to have an overview of their ability to assess the link property. We made also a statistical evaluation of estimators to measure their forecasting properties.

9.6.1.1. *Temporal assessment*

We observe a fast speed crossing wherein nodes are able to communicate within a few seconds ($4s$). Figure 9.7 shows the result of the first scenario.

Regarding the distribution of the d_r and d_f, nodes are able to communicate within a few seconds (from $6s$ to $10s$, while stochastic losses can be observed that result from a significant fading (Rayleigh) effect). The PRR computed over a history of packets declares the link disrupted at 15s, but the effective disruption occurs at 11s.

Concerning the F-LQE, Figure 9.7(b) shows the smoothed PRR (SPRR) evaluating the link quality and the link stability estimation (SF). Figure 9.7(c) shows the unidirectionality level of a link estimator (ASL). The SPRR follows the corresponding PRR (Figure 9.7) trace with a smoothing trend, but the estimator is clearly not reactive enough and detects the disruption too late. This results from the EWMA filter that provides more stability than reactivity to the estimator. The SF estimator detects that the link quality is changing at $11s$, because the link is disrupted. However, the variation indicated by the estimator does not reflect a disruption case but only a slight variation of the link quality. The disruption can be clearly detected at the end with a more important value of the indicator. Regarding the ASL indicator, the variation of d_r and d_f distribution introduces light fluctuations, indicating a low probability of having an asymmetric link.

Regarding estimators from HoPS (Figure 9.7(d) and (e)), we observed the slow convergence time of the short-term estimation; while a link is disrupted, the estimator declares the link quality as not disrupted. But the long-term estimator indicates a correct decreasing trend. Consequently, the EWMA filter is well used for the long-term estimation but is not suitable for a short-term estimation which also smooths the estimation. In the same way, the link quality trend and the variation indicators are affected by the long reactivity of the short- and long-term estimations and react too slowly when a disruption occurs.

The estimators of F-ETX are shown in Figure 9.7(f) and (g). In contrast with other LQEs, F-ETX is more reactive than the others and declares the link disrupted earlier (at 13s). The trend estimation indicates a degradation of the link quality via consecutive negative values. This is confirmed by the link stability estimator indicating a low level of stability and a decrease. We also observe that the stability estimator declares the link disrupted earlier than the link quality estimator (12*s*). Concerning the unidirectional indicator, it gives a positive value (at 10s) indicating that the link can be unidirectional.

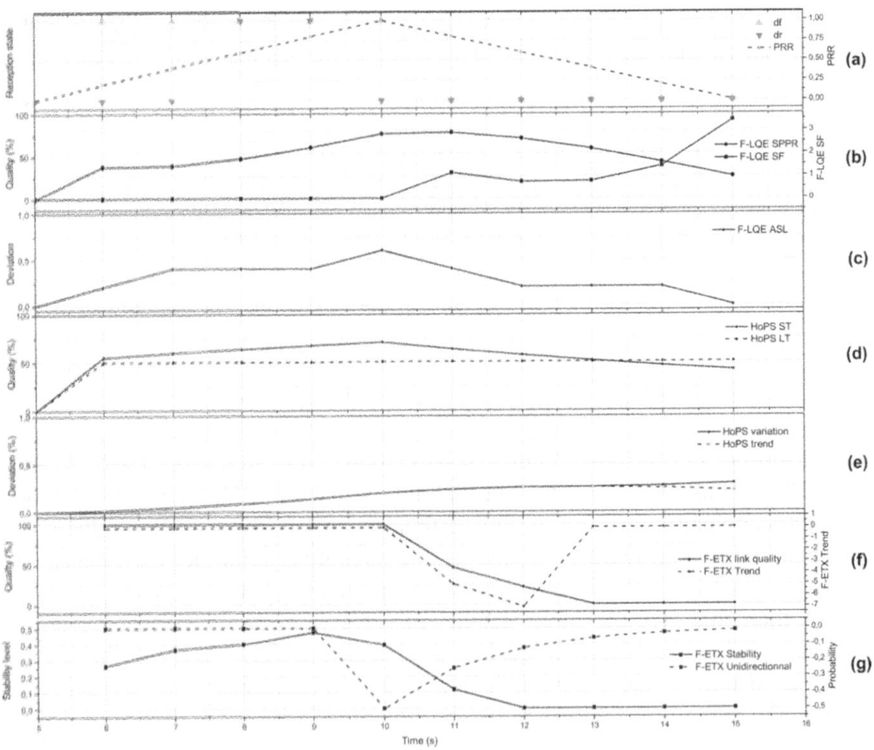

Figure 9.7. *Fast speed crossing*

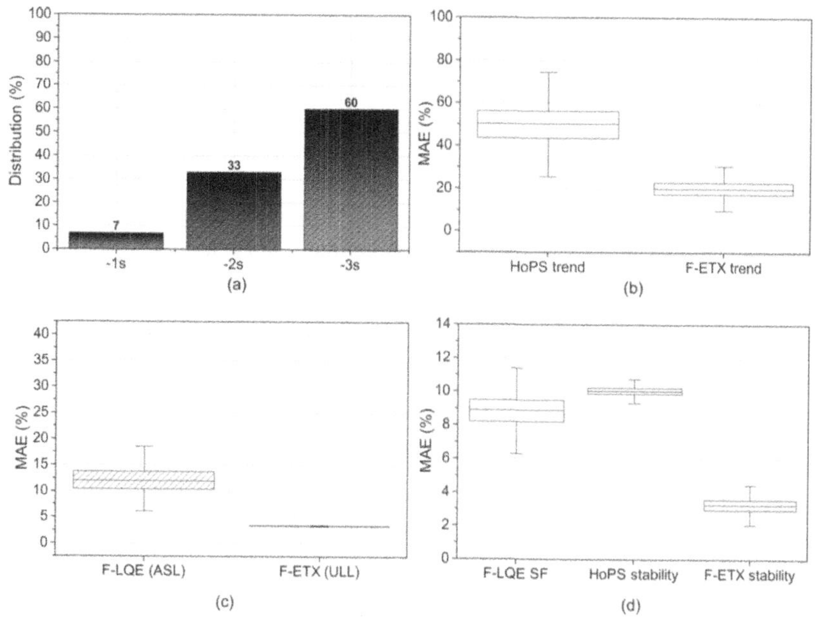

Figure 9.8. *Statistical evaluation*

9.6.1.2. *Statistical analysis*

While previous evaluations provide detail about the strength and weakness of LQEs, we extend our assessment with a statistical analysis of all the links available in the scenario.

We have observed that the link quality estimator of F-ETX is more reactive than F-LQE and HoPs. In this statistical study, we are focused on how this estimator can anticipate disruptions compared to the PRR solution based on a history of 5 packets. Figure 9.8 (a) shows that F-ETX is clearly the best solution for anticipating disruptions before the PRR solution. Since it is based on the dynamic window size, the metric is more reactive and tracks, link states change very well. In addition, F-ETX assesses both link directions, unlike a PRR solution that only evaluates the downlink.

The rest of the statistical analysis is made with the Mean Absolute Error (MAE) that measures the magnitude of the predicted estimation and the current outcome. A low score indicates a good prediction while a bigger value indicates a greater error between the prediction and the current value. The link quality trend is additional information that determines the current course of the link quality. Figure 9.8(b) shows the link quality trend of HoPs and F-ETX. We observe the better ability of our estimator to give the tendency of the link quality compared to HoPS, even if both of them are based on the link quality estimator and computed with an EWMA filter.

Their ability to track the link quality course depends on the ability of the short-term link quality estimator. HoPS-ST suffers from lag with the use of the EWMA filter impacting the HoPS-LT. On the other hand, the link quality estimator from F-ETX is reactive but unstable. That is why, with the use of EWMA, the estimation is stabilized given a better long-term estimation than the HoPS-LT overestimating the tendency.

Figure 9.8(c) shows the unidirectional link estimator of F-LQE and F-ETX. During the simulation, any effective unidirectional links are present. The ASL estimation often makes a single reading of the reception ratio of the up and downlink different when high propagation disturbances are present. Our indicator adopts another strategy based on the variation between the up and downlink. As a result our estimation is more robust to disturbances and gives more accurate information about the potential of a bidirectional link becoming unidirectional.

Tracking link stability is an essential feature for detecting and differentiating transient and persistent links. We have compared in Figure 9.8(d) the variation of the value given by these estimators to the current variation observed from the delivery and forward ratios. F-ETX estimator gives the lowest MAE compared to the others. Because the HoPS indicator only tracks the variation between the HoPs ST and LT estimations, it is not really related to the link stability. For F-LQE, the estimation is based on a PRR history generating consecutive error predictions.

9.6.2. *Performance of routing protocols*

We investigate the impact of the F-ETX metric, when it is used as the metric. We have developed a simulation model retracing the behavior of the B.A.T.M.A.N. (Better Approach To Mobile Ad-hoc Networking) protocol. We have implemented the F-ETX metric into the routing protocol and define our metric as the main. We compare the performances of our modified protocol to a couple of protocols, such as OLSR (Optimized Link State Routing Protocol) and AODV (Ad hoc On Demand Distance Vector). We have retained these protocols, because they get routing information with a different approach, proactive for OLSR and reactive for AODV. In order to rank protocols, we regard two indicators. The first one is the Packet Delivery Ratio (PDR), which indicates the number of packets successfully delivered to a destination. The second one is the end-to-end delay bringing information on the time taken by a packet to be transmitted from a source to a destination.

9.6.2.1. *Influence of the node number*

We observe the impact of the node number present in the network on the routing performances. For this purpose, nodes transmit with a constant bit rate, UDP datagrams. We analyze six traffic patterns during the simulation period, which represent a total of 2688 bytes exchanged. Figure 9.9 shows the average PDR and the average end-to-end delay.

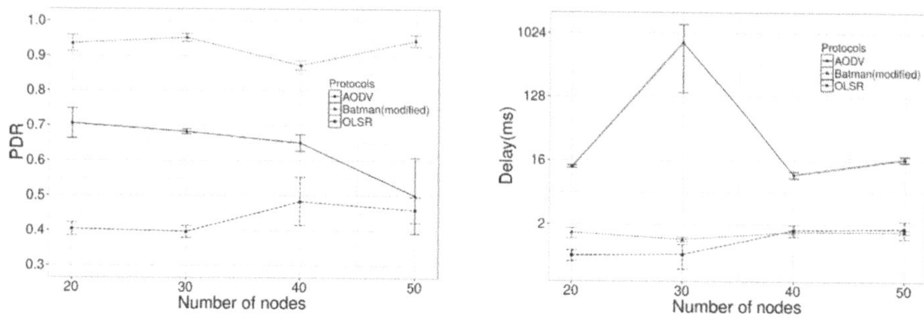

a) Average packet delivery rate b) Average end-to-end delay

Figure 9.9. *Influence of the node number*

Our modified protocol is clearly the best one, since it presents the best performance compared to OLSR and AODV. Regarding the PDR, the node number has a great influence on the two protocols, but with contrasting effects. When the node number increases, the OLSR protocol gets better performances whereas the AODV protocol obtains less efficients performances. Regarding our protocol, the node number has some impact on its performance.

Concerning the end-to-end delay, expected proactive protocols, including OLSR and our protocol, get the minimum end-to-end delay compared to the reactive protocol, AODV. Since the two proactive protocols discover the network topology periodically, nodes are able to find the best as soon as data is required to be transmitted. However, reactive protocols like AODV trigger the route discovery as soon as data have to be transmitted. That is why the end-to-end increases, since this discovery phase introduces a delay.

9.6.2.2. *Influence of the applicative throughput*

We investigate the impact of the applicative throughput on the routing performance. We study scenarios with 20 and 50 nodes and fix the applicative throughput with different sending periods, 600, 300 and 150 ms in order to differently stress the routing path. Figure 9.10 shows the resulting PDR and end-to-end delay.

Concerning the PDR, the modified version of BATMAN gets the best ratio with a PDR ≥ 0.8. Even if the throughput of the application impacts all protocols, the modified BATMAN maintains the best performance. Concerning the end-to-end delay, as expected, proactive protocols have lower delay than the reactive protocol AODV. The end-to-end delay obtained by proactive protocols is close ($< 2.7ms$). Finally, the modified BATMAN protocol appears as the best one, since it has the best PDR ratio and a low end-to-end delay close to the OLSR delay.

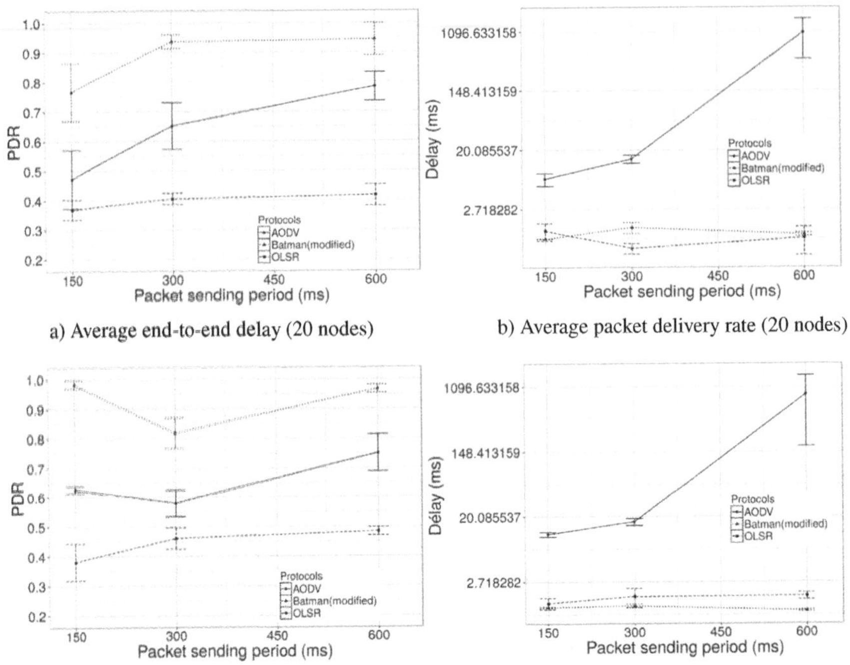

a) Average end-to-end delay (20 nodes)

b) Average packet delivery rate (20 nodes)

a) Average packet delivery rate (50 nodes)

b) Average end-to-end delay (50 nodes)

Figure 9.10. *Influence of the applicative throughput*

9.7. Conclusion

The F-ETX has been proposed to overcome the intrinsic limitations of LQEs design for WSNs. To deal with the dynamic of vehicular networks, the metric relies on a dynamic window size. F-ETX is formed of four estimators, two dedicated to the link quality assessment and another two to the link state determination. The first couple assesses both the short and long-term link quality, and the other two determine if the link is stable and unidirectional. We have developed a framework in order to integrate each estimator into the routing process. To this end, a link is assessed both on its quality and its state.

We have investigated the performance of the F-ETX metric and its impact into a routing protocol, through a realistic simulation environment. Compared to other current multi-estimator solutions, our metric is more reactive and accurate and provides the best predictions. We have implemented the metric into a proactive routing protocol (BATMAN) and compare its performance to another proactive protocol (OLSR) and a reactive protocol (OLSR). Regardless of the node number and the application throughput, our modified BATMAN obtains the best results in terms of the packet delivery ratio and has a similar delay to the OLSR protocol.

9.8. Bibliography

[BAC 10] BACCOUR N., KOUBÂA A., YOUSSEF H. *et al.*, "F-LQE: a fuzzy link quality estimator for wireless sensor networks", *Proceedings of the 7th European Conference on Wireless Sensor Networks (EWSN'10)*, pp. 240–255, 2010.

[BAC 12] BACCOUR N., KOUBÂA A., MOTTOLA L. *et al.*, "Radio link quality estimation in wireless sensor networks: a survey", *ACM Transactions on Sensor Networks*, vol. 8, no. 4, pp. 34:1–34:33, 2012.

[BAC 15] BACCOUR N., KOUBÂA A., YOUSSEF H. *et al.*, "Reliable link quality estimation in low-power wireless networks and its impact on tree-routing", *Ad Hoc Networks*, vol. 27, no. C, pp. 1–25, 2015.

[BEN 12] BENIN J., NOWATKOWSKI M., OWEN H., "Vehicular Network simulation propagation loss model parameter standardization in ns-3 and beyond", *Southeastcon, 2012 Proceedings of IEEE*, pp. 1–5, March 2012.

[BIN 15a] BINDEL S., CHAUMETTE S., HILT B., "A novel predictive link quality metric for mobile ad-hoc networks in urban contexts", *Ad Hoc Networks: 7th International Conference, AdHocHets 2015*, San Remo, pp. 134–145, September 2015.

[BIN 15b] BINDEL S., CHAUMETTE S., HILT B., "F-ETX: an enhancement of ETX metric for wireless mobile networks", *Communication Technologies for Vehicles: 8th International Workshop, Nets4Cars/Nets4Trains/Nets4Aircraft 2015*, Sousse, pp. 35–46, May 2015.

[BOA 09] BOANO C.A., VOIGT T., DUNKELS A. *et al.*, "Poster abstract: exploiting the LQI variance for rapid channel quality assessment", *International Conference on Information Processing in Sensor Networks, IPSN 2009*, pp. 369–370, April 2009.

[CER 05a] CERPA A., WONG J.L., KUANG L. *et al.*, "Statistical model of lossy links in wireless sensor networks", *IPSN 2005. Fourth International Symposium on Information Processing in Sensor Networks*, pp. 81–88, April 2005.

[CER 05b] CERPA A., WONG J.L., POTKONJAK M. *et al.*, "Temporal properties of low power wireless links: modeling and implications on multi-hop routing", *Proceedings of the 6th ACM International Symposium on Mobile Ad Hoc Networking and Computing*, MobiHoc '05, New York, pp. 414–425, 2005.

[DEC 03a] DE COUTO D.S.J., AGUAYO D., BICKET J. *et al.*, "A high-throughput path metric for multi-hop wireless routing", *Proceedings of the 9th Annual International Conference on Mobile Computing and Networking*, MobiCom '03, New York, pp. 134–146, 2003.

[DEC 03b] DE COUTO D.S.J., AGUAYO D., CHAMBERS B.A. *et al.*, "Performance of multihop wireless networks: shortest path is not enough", *ACM SIGCOMM Computer Communication Review*, vol. 33, no. 1, pp. 83–88, January 2003.

[GAB 14] GABTENI H., HILT B., DROUHIN F. *et al.*, "A novel predictive link state indicator for ad-hoc networks", *2014 IEEE Global Communications Conference*, pp. 149–154, December 2014.

[HEI 12] HEINZER P., LENDERS V., LEGENDRE F., "Fast and accurate packet delivery estimation based on DSSS chip errors", *INFOCOM, 2012 Proceedings IEEE*, pp. 2916–2920, March 2012.

[IEE 16] IEEE, "IEEE Standard for Low-Rate Wireless Personal Area Networks (WPANs)", *IEEE Std 802.15.4-2015 (Revision of IEEE Std 802.15.4-2011)*, pp. 1–709, April 2016.

[LAI 03] LAI D., MANJESHWAR A., HERRMANN F. *et al.*, "Measurement and characterization of link quality metrics in energy constrained wireless sensor networks", *Global Telecommunications Conference, 2003. GLOBECOM '03. IEEE*, vol. 1, pp. 446–452, December 2003.

[LIU 14] LIU T., CERPA A.E., "Data-driven link quality prediction using link features", *ACM Transactions on Sensor Networks*, vol. 10, no. 2, pp. 37:1–37:35, January 2014.

[QUA 11] QUARTULLI A., C.L., "Client announcement and Fast roaming in a Layer-2 mesh network", Technical Report #DISI-11-472, University of Trento, 2011.

[REN 11] RENNER C., ERNST S., WEYER C. *et al.*, "Prediction accuracy of link-quality estimators", *Wireless Sensor Networks: 8th European Conference, EWSN 2011*, Bonn, pp. 1–16, February 2011.

[RIL 10] RILEY G., HENDERSON T., "The ns-3 Network Simulator", in WEHRLE K., GÜNES M., GROSS J. (eds.), *Modeling and Tools for Network Simulation*, Springer, Berlin, 2010.

[SPU 13] SPUHLER M., LENDERS V., GIUSTINIANO D., "BLITZ: wireless link quality estimation in the dark", *Proceedings of the 10th European Conference on Wireless Sensor Networks (EWSN'13)*, pp. 99–114, 2013.

[SRI 06] SRINIVASAN K., DUTTA P., TAVAKOLI A. *et al.*, "Understanding the causes of packet delivery success and failure in dense wireless sensor networks", *Proceedings of the 4th International Conference on Embedded Networked Sensor Systems (SenSys '06)*, New York, pp. 419–420, 2006.

[SRI 10] SRINIVASAN K., DUTTA P., TAVAKOLI A. *et al.*, "An empirical study of low-power wireless", *ACM Transactions on Sensor Networks*, vol. 6, no. 2, pp. 16:1–16:49, March 2010.

[WOO 03] WOO A., CULLER D., Evaluation of efficient link reliability estimators for low-power wireless networks, Report no. UCB/CSD-03-1270, EECS Department, University of California, Berkeley, 2003.

[ZAM 07] ZAMALLOA M.Z.N., KRISHNAMACHARI B., "An analysis of unreliability and asymmetry in low-power wireless links", *ACM Transactions on Sensor Networks*, vol. 3, no. 2, 2007.

10

Autonomic Computing and VANETs: Simulation of a QoS-based Communication Model

10.1. Introduction

The complexity of intelligent transportation systems (ITSs) management is growing. This is emphasized by the use of heterogeneous technologies, which enables different kinds of communications. Adapting the Autonomic Computing paradigm to ITSs and in particular to vehicular ad hoc networks (VANETs) in order to enhance the performance of communications in such changing environments is a challenging task. This approach can be applied to improve the performance of communication protocols, such as broadcasting methods. The broadcasting communication mode is widely used in VANETs for sending emergency messages and road-traffic information or to help routing protocols to determine routes. This communication mode is known to be difficult to achieve efficiently, as it depends on the network density. Indeed, broadcasting methods may cause network congestion if they are not well designed. This chapter introduces the application of Autonomic Computing principles within VANETs' environments in order to enhance the performance of QoS-based communications thanks to the self-management concept. In such environments, the design of a QoS-based broadcasting protocol is presented as a usage case. A state of the art concerning the Autonomic Computing paradigm and its application in VANETs is first presented in section 10.2. Then, section 10.3 describes the existing broadcasting methods and protocols in wireless ad hoc networks and especially in VANETs. This state of the art helps introduce in section 10.4 the design of a QoS-based autonomic broadcasting protocol in VANETs in order to deliver messages in accordance with the given message classes and network

---

Chapter written by Nader MBAREK, Wahabou ABDOU and Benoît DARTIES.

density levels. This approach allows each vehicle to dynamically adapt its broadcasting strategy not only with respect to the network density but also according to the class of the message to be sent: emergency (high-priority), road-traffic (medium-priority) or comfort message (low-priority). Finally, in section 10.5, we present the simulation details of a QoS-based communication model in autonomic VANETs as well as the design and evaluation of a self-managed broadcasting protocol called ADM (autonomic dissemination method), which serves as an example of an autonomic broadcasting approach.

10.2. Autonomic Computing within VANETs

10.2.1. *Autonomic Computing*

Traditionally, networks and systems management is a manually controlled process. Thus, it is necessary to have a human intervention of one or several operators in order to manage all the aspects concerning the dynamic evolution of a system or a network. The creation of self-management systems with limited human interventions was the vision behind introducing autonomy within the IT environment in order to cope with the increasing complexity and excessive maintenance costs [GRO 02]. Such autonomic systems are able to be self-organized. Thus, networks become a collection of interconnected self-governed entities, where human intervention is limited to high-level directives, and system management details are transparent for the administrators.

The first initiative dealing with this paradigm is inspired by biological systems and, in particular, the autonomic nervous system. Indeed, the term Autonomic Computing is partly owed to the autonomic nervous system [HOR 01]. This management concept is based on a holistic approach, where all research fields are implicated in contributing to the evolution of a global autonomy in networks.

Although the objectives list of the self-management concept was extended after 2001 (year of birth of this new paradigm), the main objectives for autonomic systems are self-configuring, self-healing, self-optimizing and self-protecting [GAN 03]. To achieve these objectives, autonomic systems have a detailed knowledge of their internal state as well as their environment [STE 03], using a continuous monitoring approach to detect eventual changes that could affect their components. Detecting changes induces the autonomic system to adjust its resources, and the monitoring continues to determine if the new measures satisfy the desired performance. That is the closed control loop of self-management systems. It enables autonomic systems to make adequate decisions while conforming to global objectives without human intervention, thanks to measurement data collected from its resources. This closed control loop is implemented by autonomic managers, which control managed resources using the manageability interfaces of sensors and effectors [IBM 05].

10.2.2. *Autonomic vehicular communications*

Adapting the Autonomic Computing paradigm to transportation systems and in particular to VANET networks in order to enhance the performance of communications within such a changing environment is a challenging task. In Hsu *et al.* [HSU 10] and Li *et al.* [LI 12], the authors describe the corresponding challenges, approaches and solutions in ITS. Indeed, they introduce the cooperative communication concept in vehicular networks. These networks should be self-managed thanks to a self-configuration function using decision elements and control loops. Monitoring and policing information will be used within cooperative VANET communications in conformance with the Autonomic Computing concepts in order to enhance vehicle safety.

The research work presented in Wodczak [WOD 12] describes the self-management capability of vehicles in order to perform autonomic cooperative communications and routing within VANETs. The author presents the architecture of an autonomic cooperative node (i.e. vehicle) on the basis of the Generic Autonomic Network Architecture (GANA). This autonomic node includes different decision elements (DE) controlling a managed entity (ME) in order to enhance the performance of Vehicle-to-Vehicle (V2V) communications.

Research challenges concerning inter-vehicular communication (IVC) are presented by Dressler [DRE 11] in four areas. The area dealing with IVC communication principles and patterns discusses the emerging IVC applications such as safety traffic and describes how V2V communications could be used for self-organized traffic control. Insaurralde [INS 12] introduces the autonomic management of autonomous underwater vehicles (AUVs) in order to provide these vehicles with self-maintenance during their missions. An autonomic AUV control architecture is proposed. The objective of this architecture is to achieve the self-management capabilities described by the Autonomic Computing paradigm.

10.3. Broadcasting protocols for VANETs

Broadcasting consists of sending a message from one node to all other nodes within a network. In wireless networks, the coverage area limits of each node restrict the propagation of the radio signal to the nodes located within the transmitter coverage area. In VANETs, wide dissemination of messages can only be ensured if some nodes relay the packets they receive. Moreover, the fact that nodes share the radio channel requires designing broadcasting strategies that minimize the risk of interference. This can be achieved by reducing the number of relays in high-density networks. This reduction should not lead to the interruption of message propagation.

Finding a good broadcasting strategy is complex in VANETs (in wireless ad hoc networks in general) because the decision of whether or not to relay each message is taken in a decentralized way. This means that none of the nodes have information on the overall network topology. Each decision is taken according to local information.

In the literature, there exist several classifications of broadcasting methods for wireless ad hoc networks [WIL 02, WU 03, STO 04]. Williams and Camp [WIL 02] classify broadcasting methods according to the information used for decision-making (probability, number of redundant copies, neighbor list). Wu and Lou [WU 03] proposed two taxonomies. The first one is based on the type of algorithm used: probabilistic or deterministic. The second classification is based on the amount of state information used in the algorithm: global, quasi-global, quasi-local and local. According to Stojmenovic and Wu [STO 04], broadcasting schemes can be classified using a taxonomy that consists of five categories: determinism, network information, reliability, Hello message content and broadcast message content. In this chapter, the classification of Williams and Camp is used. The broadcasting families are grouped in two main categories: deterministic and stochastic methods (see Figure 10.1).

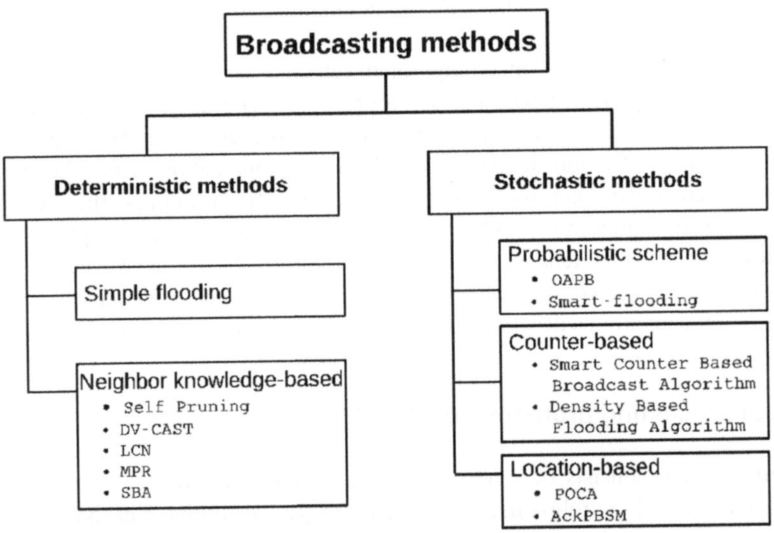

Figure 10.1. *Classification of broadcasting methods*

10.3.1. *Deterministic methods*

A broadcasting method is deterministic if its process is predictable. This group includes simple flooding and neighbor knowledge-based methods.

10.3.1.1. *Simple flooding*

Simple flooding is the simplest broadcasting method. Every packet is relayed exactly once by each node. Any redundant copy of the packet received later is ignored. Thus, in a network consisting of n nodes, n copies of the packet will be sent. The drawback of this method is that it does not take into account the network density. In high-density networks, the Simple flooding algorithm would generate many redundant copies of broadcasted packets, leading to the overuse of the radio resources.

10.3.1.2. *Neighbor knowledge-based methods*

Neighbor knowledge-based methods compare neighbor lists before relaying packets. Nodes exchange Hello packets in order to discover the local network topology and to build up their neighbor lists.

Flooding with Self-Pruning [LIM 00] uses a one-hop neighbor list. This list is inserted into the broadcast packets. This allows each receiver to compare its own list to the one included in this packet. If the lists are identical, the packet is dropped, otherwise the packet is relayed. Other methods such as Distributed Vehicular Broadcast (DV-CAST) [TON 10] and least common neighbor (LCN) [YU 06] also rely on one-hop neighbor lists.

The multi-point relay (MPR) technique [NGU 07] is a neighbor knowledge-based broadcasting method. To reduce the number of redundant packets in the network, each node chooses several nodes among its neighbors that will relay its communications. The selected nodes are called MPRs. The MPRs are selected among the one-hop neighbors so that they enable reaching all two-hop neighbors. The goal is to have the smallest list of MPRs in the network, which optimizes communications. This method requires a bidirectional link. When a node sends a packet, all of its neighbors receive it, but only the MPRs of the source node will relay the message. Thus, each node will have a list of all the nodes that have chosen it as a relay (MPRs selectors' list).

Scalable Broadcast Algorithm (SBA) [PEN 00] uses topology information within two hops. When a node, which runs SBA, receives a broadcast packet, it uses the source identifier (the node that originates the packet or the last relay node) and its neighbor list to determine all additional nodes (N) in its neighborhood that would receive the packet if it is relayed. If N is empty, the packet is dropped. Otherwise, the node sets a timeout called RAD (Random Assessment Delay). If it receives another copy of the same packet, N is recalculated. On expiry of the RAD timer, the packet is relayed if N is not empty. To optimize this algorithm, the authors advise calculating the waiting time (t) so that nodes with many neighbors are prioritized (see equation 10.1):

$$t = \frac{N_{\max}}{n} \qquad [10.1]$$

where n is the number of neighbors of the receiver and N_{max} is the maximum "n" of the receiver's neighbors.

For static or low mobility networks, neighbor knowledge-based methods can achieve good performance. However, in high-mobility networks, like VANETs, information about the neighbors becomes inaccurate rapidly. Thus, this family of methods is hardly applicable for vehicular networks.

10.3.2. *Stochastic methods*

The stochastic methods statistically assess the gain that could be obtained if the packets are relayed by a given node. They include probabilistic scheme, counter-based and location-based methods.

10.3.2.1. *Probabilistic methods*

To avoid the broadcast storm problem [NI 99] and adjust broadcasting strategies depending on the network density, probabilistic methods mainly use a parameter that serves to relay (or not relay) each received packet. In fact, for a given network density, there exists p_s, a probability threshold value ($0 \leq p_s \leq 1$), which would allow all nodes to receive the packets, reducing the number of unnecessary repetitions and leading to few collisions. Any other value $p > p_s$ would not lead to better coverage, but may downgrade the quality of the communication. As p_s varies locally in the network, the main challenge of the probabilistic methods is to determine its correct value. Some approaches to dynamically assign value to p_s are proposed in the literature. They combine probabilistic methods with some other techniques for assessing the network density (e.g. counter-based or distance-based methods).

Optimistic Adaptive Probabilistic Broadcast (OAPB) [ALS 05] adapts the probability of each vehicle according to its number of neighbors within two hops. This allows the protocol to adjust the broadcasting strategies to local densities. The neighborhood of each vehicle is discovered thanks to Hello packets. Smart-flooding

[ABD 12a] also aims to adapt the broadcasting probability to the local density. In addition to the probability, this protocol introduces several other parameters such as the number of retransmissions for each packet and the delay between successive retransmissions. To achieve good tuning of these parameters for various density levels, the authors have used a genetic algorithm. It is important to note that Smart-flooding does not use Hello packets to evaluate the local density. It takes advantage of traditional exchanges between nodes to estimate the number of neighbors for each node.

10.3.2.2. *Counter-based methods*

The principle of the counter-based methods is simple: the more copies of the same packet a node receives, the less likely that it is useful to relay this packet. Upon reception of the first copy, the node initializes a counter C to 1 and sets a timeout RAD (Random Access Delay). During the waiting period, C is incremented upon reception of a new copy of the packet. When the RAD expires, C is compared to a threshold value, C_t. If $C < C_t$, the packet is broadcast, otherwise it is dropped. Like probabilistic methods, one challenge is to find an appropriate value for C_t. Ni *et al.* [NI 99] demonstrated that the additional area covered by the broadcasting process decreases significantly when the number of redundant copies increases.

Bani Yassein *et al.* [BAN 07] proposed the Smart Counter-Based Broadcast Algorithm that adapts C_t according to the network density. Thanks to Hello packets, the nodes build neighbor lists. The size of these lists allows dynamically adjusting C_t. Karthikeyan *et al.* [KAR 10] introduced a method named Density-Based Flooding Algorithm. This method defines two categories of nodes according to their number of neighbors with respect to a given threshold, τ. Each node decides to relay each packet depending on its own category and the one of the packet's last hop.

10.3.2.3. *Location-based methods*

Before relaying a message in the context of location-based methods, the node evaluates the additional coverage area that will result from this retransmission. These methods do not consider whether nodes exist within that additional area or not. AckPBSM [ROS 09] and POCA [NAN 10] use this approach and set lower RAD to nodes that are far from the source node (or last-hop relay node). To evaluate the extra coverage area, the node can use the distance between itself and each node that has previously relayed the message (distance-based scheme) or the geographical coordinates (location-based scheme). In both distance-based and location-based schemes, a RAD timeout is set, and the message is relayed if the additional coverage area is higher than a fixed threshold.

10.4. Autonomic broadcasting within VANETs

After a description of the Autonomic Computing paradigm and some applications of its concepts within VANETs, we detailed the state of the art concerning the use of broadcasting protocols within VANETs. In the following sections, we present a study concerning the application of Autonomic Computing concepts to an example of these broadcasting protocols within VANETs. Thus, a self-management architecture is specified to enable QoS-based autonomic broadcasting while demonstrating that such kind of broadcasting is an optimization problem in VANETs' environment.

10.4.1. *Optimization of broadcasting protocols in VANETs*

Designing an efficient broadcasting protocol requires meeting several objectives that could be antagonistic: for instance, transmitting messages to the maximum of nodes while avoiding the overuse of the radio channel and delivering packets as quickly as possible, knowing that this speed may cause radio interferences. In a nutshell, this is clearly a multi-objective optimization problem for which each solution is a set of parameters that defines a broadcasting strategy. Depending on the protocol to optimize, the parameters could be a probability, the boundaries of the RAD, some thresholds, etc.

In [ABD 12a], the authors define a broadcasting strategy as a set of four parameters:

– the probability of relaying a packet (*P*). It is inherited from the classical probabilistic methods. When a node receives the first copy of a broadcast packet, it decides to relay it or not, depending on the value of *P*. The following three parameters are applicable only if the node decides to relay the packet.

– the number of repetitions (*Nr*). In low-density networks, when a node broadcasts a packet, it is not unusual that it has no neighbor in its coverage area that will receive the message and relay it. Sending the packet several times, particularly in a context of mobility, the node increases the chance that the packet will be received and relayed. *Nr* is also useful when the first transmitted packet is lost due to a collision or poor radio propagation quality.

– the delay between two successive repetitions (*Dr*). When a node transmits the same packet several times ($Nr > 1$), it is important to determine the frequency with which the copies of the same packet will be transmitted. A very short delay could result in many collisions, whereas a very long delay may slow down the broadcast.

– the packet's lifetime. It allows a limited spread of packets within the network and/or for a long period of time. The maximum number of hops allowed for each packet, *TTL* (Time To Live), could be used in the context of broadcasting protocols. Geographical coordinates or transmission time can replace this parameter.

These parameters allowed the authors to tune their protocol named Smart-flooding. The optimization process of the broadcasting strategies defined by these parameters (*P, Nr, Dr* and *TTL*) was carried out using four criteria:

– the average Number of Collisions (*NC*);

– the Propagation Time (*PT*). This is the time between the transmission of a packet and its reception by all nodes of the studied area;

– the total number of Retransmissions during the simulation (*R*);

– the Full Reception ratio (*FR*). This refers to the guarantee that the broadcast packets will be received by all nodes (the reachability). A simulation in which all nodes receive the packet is considered successful. On the contrary, if the network conditions (propagation or topology) do not allow the reception of the packet by all nodes, the simulation is considered as a failure. *FR* is the ratio of the number of successes on the total number of repetitions of each scenario executions.

The first three criteria are to be minimized, while *FR* is to be maximized. *NC* and *R* enable measuring the radio channel usage: high values indicate that the evaluated strategy is likely to interfere with other communications in the network. The calculation of *NC*, *PT* and *R* takes into account successful simulations only.

10.4.2. *Self-management architecture*

To be efficient, broadcasting protocols in VANETs should adapt their communication strategies according to not only the network density but also the priority level of the message that has to be disseminated. Such protocols can be specified using the closed control loop implemented by an autonomic manager within a mobile node (vehicle). The latter is considered a managed resource according to Autonomic Computing concepts presented in section 10.2.1. The resulting architecture, enabling broadcasting strategy optimization according to VANETs' environment characteristics and change occurrence, is detailed below.

A self-management approach of radio communications ensures the robustness of a broadcasting protocol. Indeed, each node (i.e. vehicle) is considered to be an autonomic element thanks to an autonomic manager that enables broadcasting decision-making according to the message priority level and takes into account environment changes in terms of density level (an example is given in section 10.5.1).

To achieve those goals, the autonomic manager implements the MAPE-K closed control loop (see Figure 10.2) and communicates with the Managed Element mobile node using Sensors and Effectors manageability interfaces.

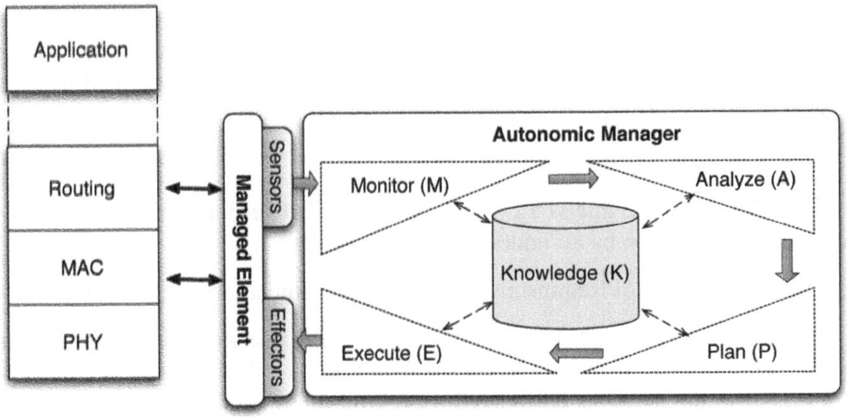

Figure 10.2. *Autonomic manager closed control loop*

Each autonomic node within a VANET continuously monitors its environment and network traffic by listening to the radio channel and provides the Monitor function (M) of the Autonomic Manager with network traffic information thanks to the Sensors manageability interface. In the self-management architecture presented in Figure 10.2, the Knowledge base should contain efficient parameters in order to enable the corresponding broadcasting strategies.

In the context of a broadcasting protocol, the Monitor determines if the received packet is a broadcasting one thanks to its destination address. In the case of a broadcasting message, the Monitor provides the Analyze function (A) with this information to follow the control loop process.

The Analyze function determines not only the priority level of the message according to the packet header information but also the current density level of the node environment (which can be evaluated using Hello packets or data packet as in the case of Smart-flooding). This information could be stored in the Knowledge base (K) of an autonomic manager.

After the evaluation of the density level, the Plan function (P) will use the density and priority values provided by the Analyze function to retrieve the adequate broadcasting strategy from the Knowledge base (K) thanks to the strategy table created by the optimization offline phase (see section 10.5.1.3). Then, the Plan function will provide the Execute function (E) with the broadcasting parameters

(*P, Nr, Dr* and *TTL* in case of Smart-flooding or ADM described in section 10.5.1) in order to change the behavior of the mobile node-managed resource by executing the corresponding actions of broadcasting strategy thanks to the Effectors manageability interface.

The self-management architecture enables the Autonomic Manager to determine how to adapt the broadcasting strategy based on the information reported by the Sensors manageability interface. Each of the four functions (MAPE) corresponding to the Autonomic Manager has a specific role; however, all share the same Knowledge base. The latter contains a set of broadcasting strategies optimized for various contexts corresponding to different density and priority levels that we describe in the following section dealing with QoS-based broadcasting.

10.4.3. *QoS-based broadcasting*

Several recent research works in VANETs and their applications highlight the need to classify messages into several classes [VEG 13, KAM 10, SUT 07]. Processing these messages depends on many criteria, such as their emergency level, their impact on the road-traffic management or the desired reachability.

In this context, three message classes can be defined for broadcast operations in VANETs (corresponding to three priority levels). Each class should satisfy a broadcast policy. These classes may be based on a single or a dual objective and may also consider other broadcast characteristics, that is, the covered nodes ratio evolution over time. These policies mainly illustrate the adaptability of a broadcasting protocol to the message contents and hence can be easily redefined or extended with additional classes.

High-level priority messages (HL for short) correspond to emergency messages, for example, safety message or accident detection. They have to be delivered as quickly as possible as they may require a prompt reaction from the driver. For these messages, a broadcasting protocol tries to minimize the required propagation time so that vehicles that are close to the broadcast source may receive the message with a very short delay. Indeed, safety messaging is a near-space application where vehicles in close proximity exchange information to increase safety awareness [VEG 13]. These applications have strict latency constraints. In addition to the reduction of the propagation, the autonomic broadcasting method will try to maximize the full reception ratio.

Medium-level priority messages (ML) correspond to road-traffic messages, for example, traffic jam report. They suppose less critical information, where driving reflexes are not part of the equation and only attention is required. These messages should cover a high ratio of nodes, while the broadcast operation requires reducing the number of radio interferences. According to [VEG 13], traffic monitoring applications require gathering information from vehicles that span multiple kilometers.

Low-level priority messages (LL) correspond to comfort messages, for example, weather information, tourist attraction or points of interest. They are optional messages whose delivery must not alter the dissemination of emergency and alert messages. The use of the radio resources has to be optimized, by reducing the number of collisions as well as the number of retransmissions, for an acceptable node coverage ratio.

Table 10.1 summarizes the classes considered in this study.

	Examples	Strategies
High-Level (HL)	Accident reports	Minimize the propagation time Maximize the reachability
Medium-Level (ML)	Traffic reports	Maximize the reachability Minimize the interferences
Low-Level (LL)	Tourist attractions	Minimize the number of collisions Minimize the number of retransmissions

Table 10.1. *Message priority levels*

10.5. Simulation of a QoS-based communication model

In this section, we first introduce a broadcasting protocol that is inspired from Autonomic Computing paradigm. Thereafter, we present the results of that protocol.

10.5.1. *ADM: autonomic dissemination method*

10.5.1.1. *Overview*

The ADM is an extension of the Smart-flooding protocol [ABD 12a]. ADM is an autonomic robust broadcasting method, which adapts its broadcasting strategies with respect to the network density and message priority level. Its architecture is described in Figure 10.3.

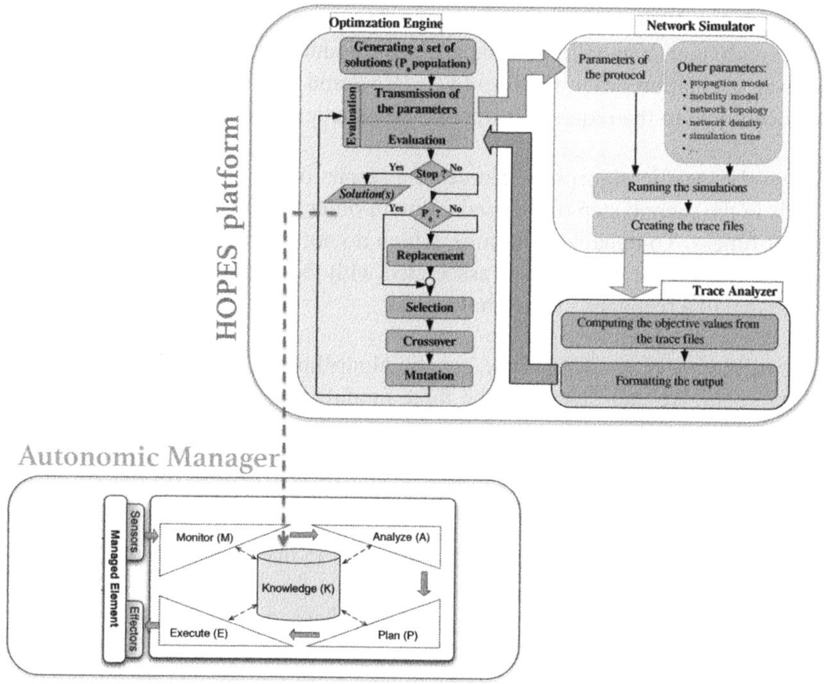

Figure 10.3. *Global flowchart of ADM. For a color version of this figure, see www.iste.co.uk/hilt/transportation.zip*

ADM relies on an offline optimization process that aims to supply the autonomic manager's Knowledge base module with good broadcasting strategies. Indeed, we optimize the parameters *P, Nr, Dr* and *TTL* using an approach that combines an optimizer, a network simulator and a trace analyzer. Figure 10.3 illustrates the interaction of these three modules. *P, Nr, Dr* and *TTL* are optimized using HOPES (Hybrid Optimization Platform using Evolutionary Algorithms and Simulations) [ABD 12a]. HOPES combines an optimizer, a network simulator and a trace analyzer.

The optimizer used is our proposed genetic algorithm aGAME [ABD 12b]. The decision variables of the problem are *P, Nr, Dr* and *TTL*. They are the different genes defining a solution (a broadcast strategy). The genetic algorithm is used to traverse the search space effectively. The optimization process starts with the random generation of the initial population (P0). The evaluation stage is split into two steps: the first one performed by the network simulator and the second one by the genetic algorithm.

Broadcast parameters (*P, Nr, Dr* and *TTL*) are transmitted to the network simulator, which integrates them with other parameters in order to better reproduce the conditions of the evaluated network. The trace files generated during the simulation are then transmitted to the analyzer module. It parses the files in order to extract the evaluation criteria values (*NC, PT, R* and *FR*) and presents the obtained results according to the required format of the genetic algorithm.

When the genetic algorithm receives the results of the trace analyzer module, it proceeds to the second step of the evaluation in order to classify solutions and assign values of fitness. To penalize solutions, which do not guarantee the full reception of transmitted packets, a constraint is associated with the problem: *FR* must be greater than or equal to a reachability threshold (*FRs*).

The remaining operations of the genetic algorithm are performed independently of the problem. At each iteration, the three modules are involved in the evaluation task. The second test of the optimization module denoted by "P_0?" checks whether the current population is the initial population. The overall optimization process leads to a set of non-dominated solutions corresponding to dissemination strategies adapted to the considered network density. This study is repeated for several densities by changing the corresponding parameter in the simulation module.

From the results of this offline optimization phase, a broadcasting protocol such as ADM builds a knowledge base that establishes a correspondence between density levels and broadcasting strategies. Density levels are represented by the number of neighboring intervals. Each node can therefore choose, depending on the density of the network in which it is located, the appropriate dissemination strategy. Then, depending on the probability of retransmission associated with the chosen strategy, the node decides whether or not to relay the packet. If the decision is to relay the communication, it applies the other corresponding parameters (*Nr, Dr* and *TTL*).

10.5.1.2. *Density evaluation*

In classical approaches, the density around a node i is often calculated by counting the number of nodes (N_i) located within the coverage area of i. These methods are based on the assumption that all nodes have uniform and identical coverage areas. This is usually the case when the radio propagation model is deterministic, such as free-space or two-ray ground reflection. However, for a more realistic model, where packet losses are distributed according to the distance between the transmitter and receiver, this definition is impractical. ADM evaluates the local density for each autonomic node on the basis of the number of active neighbors from which it received the packets. During communication, each node builds a view of its neighborhood. This view depends on the neighbor list having transmitted or relayed packets. Each autonomic node maintains a history in which it associates with each received packet a list of nodes having sent or relayed it. Upon reception the first copy of a packet, its identifier and the source/relay address are recorded within the autonomic manager Knowledge base in a table called local view.

When a redundant copy is received, the address of the new relay is appended to the local view table list of addresses (L) corresponding to the packet. Each address is recorded only once for each packet; hence, receiving multiple copies issued by one neighbor does not lengthen the list of addresses for the concerned packet. When the table is full, the oldest information is replaced by the new one according to the FIFO (*First In, First Out*) principle. The current number of neighbors (N_i) for each autonomic node i is equal to the average number of transmitters for all the packets stored in L (see equation 10.2):

$$N_i = \frac{\sum_{j=1}^{n} | L(j) |}{n} \qquad [10.2]$$

where n is the total number of packets in the local view table and $|L(j)|$ is the number of nodes that issued/relayed the jth packet in the table.

10.5.1.3. *Calibration*

We run the optimization process in order to find good broadcasting strategies that will be used by ADM for various network density levels. In this section, we show values for four density levels. We consider as a topology model a convoy of vehicles lined up for 10 km. To illustrate different density levels, we varied the inter-vehicle distance. Table 10.2 shows the parameters of the topology used for different density levels.

As in most multi-objective problems, the optimization process returns as a result of several potential solutions that offer a compromise between the different objective functions (*NC, PT, R, FR*). To refine the obtained results, we used a multiple-criteria decision-making approach based on preferences.

Density level	Number of vehicles	Inter-vehicle distance	Average number of neighbors
High (Urban)	400	25 m	26
Medium (Suburban)	134	75 m	10
Low (Highway)	50	200 m	5
Very low (Rural)	10	1000 m	1

Table 10.2. *Topology parameters for different network density levels*

For sending high-priority messages, we select the solution that allows delivering packets as quickly as possible while covering the largest number of nodes in the network. For medium-priority messages, the first criterion taken into account is reachability (*FR*). Then, among the solutions that have an *FR* value almost equal to 1 (the maximum), we select the one which causes the least collision. Finally, for low-priority messages, the goal is to send packets while slightly using the wireless channel. The first and second criteria are, respectively, *NC* and *R*. The broadcasting parameters for the three priority levels and the objective functions values corresponding to various density levels are presented in Tables 10.3–10.6, respectively, for high-, medium-, low- and very low-density networks. For each scenario, we use one source node located at the end of the convoy of vehicles. Scenarios with multiple source nodes are discussed in section 10.5.3.

Message classes	Broadcasting parameters				Performance results			
	P	Nr	Dr	TTL	NC	PT	R	FR
HL	0.329	1		32	497	0.051	131	99.6%
ML	0.258	2	1.721	15	347	0.106	207	100%
LL	0.188	1		39	190	0.048	75	86.8%

Table 10.3. *ADM's parameters and performance results for a high-density network*

Message classes	Broadcasting parameters				Performance results			
	P	Nr	Dr	TTL	NC	PT	R	FR
HL	0.776	1		26	166	0.044	104	100%
ML	0.519	2	0.951	16	93	0.121	139	100%
LL	0.291	2	0.276	27	35	0.209	82	75.8%

Table 10.4. *ADM's parameters and performance results for a medium-density network*

In high-density networks, the probability of relaying the packets is low (see Table 10.3). When $Nr = 1$, the *Dr* cell (the delay between successive repetitions) has been shaded, as this parameter is only applicable when $Nr > 1$. For high-priority messages (in a high-density network), relaying each packet only once, a probability of about 0.3, allows rapid dissemination of the message. However, this probability value generates a large number of collisions. This drawback is mended for medium-priority messages. To reduce the number of collisions and increase the reachability (*FR*), we selected a solution with a lower probability and number of repetitions equal to 2. Moreover, as the repetitions are not made in burst, the risk of interference is reduced.

For low-priority messages, it is worth noting that the results only concern the packets that have been received by all vehicles. In other words, 86.8% of packets that are received spread quickly (due to low competition in the access to the radio channel), but 13.2% of them are not completely delivered.

Following the same reasoning, we obtain the broadcasting parameters for suburban and highway scenarios (Tables 10.4 and 10.5, respectively).

For the scenario of the rural area, the low-density level of the network implies the need to retransmit each packet many times (see Table 10.6). Indeed, in this scenario, VANETs behave like delay tolerant networks (DTNs) [PAR 12]. In such a context, since the radio channel is rarely used, even if ADM is able to differentiate broadcasting strategies according to the message class, in practice, these classes scarcely impact the communication process. The main constraints that must be met are: obtaining a probability (*P*) close to 1 and a high number of repetitions (*Nr*).

Message classes	Broadcasting parameters				Performance results			
	P	Nr	Dr	TTL	NC	PT	R	FR
HL	0.999	4	1.147	40	31	0.092	199	100%
ML	0.916	2	0.729	28	24	0.124	90	100%
LL	0.649	2	1.933	34	10	1.414	66	82.8%

Table 10.5. *ADM's parameters and performance results for a low-density network*

Message classes	Broadcasting parameters				Performance results			
	P	Nr	Dr	TTL	NC	PT	R	FR
HL	0.833	28	0.233	28	58	13.09	1167	99.8%
ML	0.896	25	1.468	34	16	28.30	1124	100%
LL	0.902	8	1.622	19	4	30.96	362	92.6%

Table 10.6. *ADM's parameters and performance results for a very low-density network*

10.5.2. *Simulation environment*

This section describes the simulation parameters that are used to compare ADM with some broadcasting methods in the literature.

The simulations were carried out using the ns-2 network simulator (2.34 version) with the Shadowing Pattern propagation model [DHO 06]. It is a realistic and probabilistic propagation model, which can produce distributions of statistical errors, such as slow and fast fading, while being easy enough to be carried out on medium to large simulations.

Figure 10.4 shows the simulated network topology, which consists of three main areas. The first zone is the main road where the average speed is 130 km/h. In the second area, the average speed is 90 km/h. Finally, the third area tallies with an urban network, where the average speed is 50 km/h. These speeds correspond to the maximum speed in France, respectively, on highways, on back roads and in urban areas. We used a mobility model that redirects vehicles at every intersection to maintain the average density (average number of neighbors) required in each area (see Figure 10.4). In addition, the low velocity within the third zone leads to an increase in the density in this part of the network.

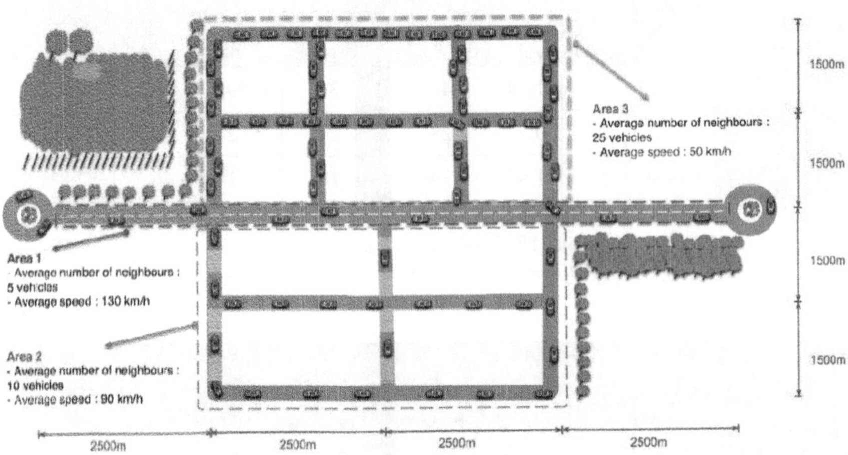

Figure 10.4. *Network topology. For a color version of this figure, see www.iste.co.uk/hilt/transportation.zip*

For these experiments, we simulated a network consisting of 600 vehicles. The simulation duration is set to 10 min. This duration allows each vehicle to move across areas and therefore to change density levels.

Packets are sent every 5 s. This allows evaluating the robustness of ADM with respect to the network traffic. At each sending phase, there is a concurrent access to the radio channel because there are several source vehicles (between 3 and 30 sources, depending on the scenario).

10.5.3. *Performance evaluation*

We evaluate the performance of ADM in a network where the density varies according to geographical locations. The aim is to assess the ability of the ADM to adapt to density changes thanks to different broadcast strategies provided by the Knowledge base of the corresponding Autonomic Manager (see Figure 10.3).

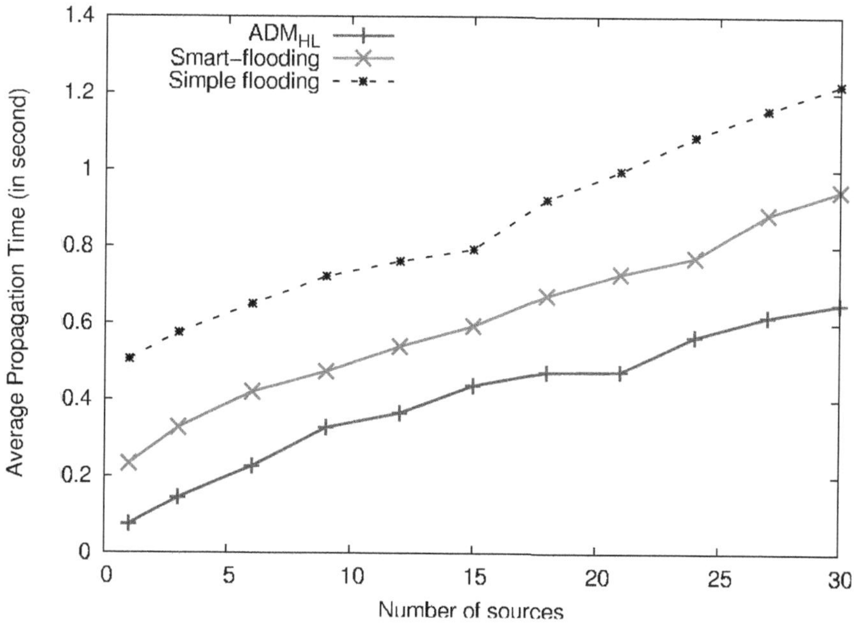

Figure 10.5. *Propagation time. For a color version of this figure, see www.iste.co.uk/hilt/transportation.zip*

The performances of communication protocols in mobile and heterogeneous density networks depend on their ability to dynamically adapt to changes in their environment. The results in Figures 10.5–10.7 clearly show that the lack of an adaptation mechanism to the density level leads to a poor performance of the Simple flooding method. Its propagation time when there are more than 18 simultaneous source nodes is at least 1 s (see Figure 10.5). This delay can be detrimental for emergency messages. Moreover, we can observe that in case of concurrent access to the radio channel, Simple flooding is struggling to deliver packets across the network (see Figure 10.6). This low reachability ratio is due to the collisions caused by redundant packets, especially in high-density areas (see Figure 10.7).

Regarding the two protocols that are able to adapt to the density, we observe that ADM has better performance results than Smart-flooding. These differences are due to the fact that Smart-flooding underestimates the network density by using a theoretical approach. ADM is not only based on this theory, but also uses experimental results.

In general, ADM delivers emergency packets in less than 700 ms in a relatively large area (even with 30 source nodes). This allows being in conformance with the limits of drivers' reaction upon alerts. Besides, for medium-priority messages (for instance, road-traffic regulation), which should be received by a maximum of nodes, ADM has a reachability ratio of almost 75%, while Smart-flooding has 66% and Simple flooding 53% in the scenario with 30 sources.

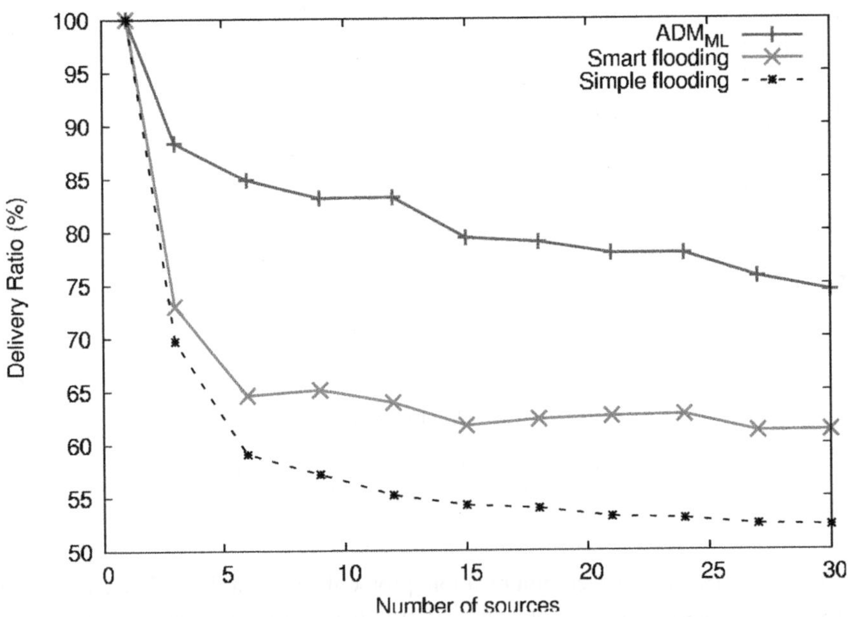

Figure 10.6. *Delivery ratio. For a color version of this figure, see www.iste.co.uk/hilt/transportation.zip*

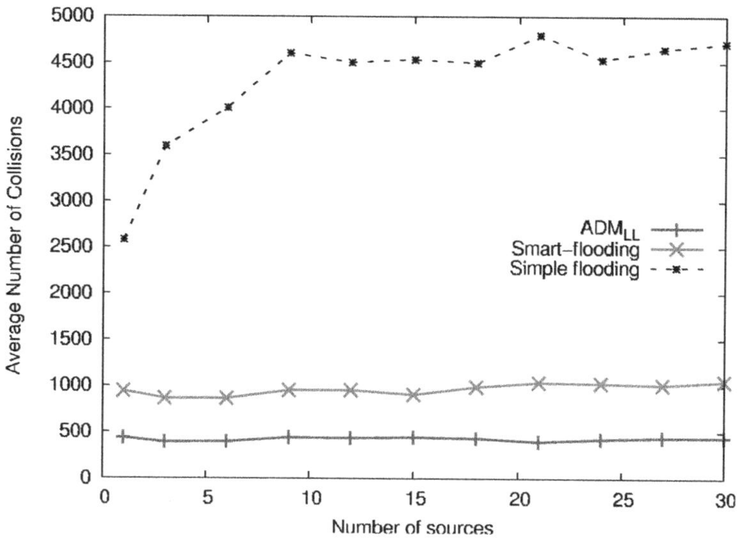

Figure 10.7. *Collisions. For a color version of this figure, see www.iste.co.uk/hilt/transportation.zip*

10.6. Conclusion

This chapter presented an application of the Autonomic Computing paradigm in VANET's communication. This approach allows building robust protocols. In this context, the design of ADM is detailed as a usage case of self-management concepts in VANETs' environment thanks to the adaptation of this method and the specification of an autonomic QoS-based broadcasting protocol.

ADM uses the obtained pre-computed broadcasting strategies thanks to an evolutionary algorithm. Each node is able to dynamically adapt its own broadcast parameters to the network density and to the message class corresponding to a priority level. The results of the simulations carried out on both homogeneous and heterogeneous density-level networks show that ADM outperforms two other broadcasting methods: the Smart-flooding protocol and the Simple flooding method. These results also reveal the scalability of ADM when the number of simultaneous transmissions significantly increases while using different message classes. Despite only considering three message classes, ADM can be easily adapted to include other message classes. These new classes will enable different features and characteristics to take into account other VANETs communication usage and interactions with the infrastructure.

10.7. Bibliography

[ABD 12a] ABDOU W., BLOCH C., CHARLET D. *et al.*, "Designing smart adaptive flooding in MANET using evolutionary algorithm", *Mobile Wireless Middleware, Operating Systems, and Applications: 4th International ICST Conference*, pp. 71–84, 2012.

[ABD 12b] ABDOU W., BLOCH C., CHARLET D. *et al.*, "Adaptive multi-objective genetic algorithm using multi-paretoranking", *14th International Genetic and Evolutionary Computation Conference*, Philadelphia, PA, pp. 449–456, 2012.

[ALS 05] ALSHAER H., HORLAIT E., "An optimized adaptive broadcast scheme for inter-vehicle communication", *2005 IEEE 61st Vehicular Technology Conference*, vol. 5, pp. 2840–2844, 2005.

[BAN 07] BANI YASSEIN M., AL-HUMOUD S., OULD KHAOUA M. *et al.*, "New Counter Based Broadcast Scheme Using Local Neighborhood Information in MANETs", University of Glasgow, Department of Computing Science, 2007.

[DHO 06] DHOUTAUT D., REGIS A., SPIES F., "Impact of radio propagation models in vehicular ad hoc networks simulations", *VANET '06 Proceedings of the 3rd International Workshop on Vehicular ad hoc Networks*, pp. 40–49, 2006.

[DRE 11] DRESSLER F., KARGL F., OTT J. *et al.*, "Research challenges in intervehicular communication: lessons of the 2010 Dagstuhl seminar", *IEEE Communications Magazine*, vol. 49, no. 5, pp. 158–164, 2011.

[GAN 03] GANEK A.G., CORBI T.A., "The dawning of the autonomic computing era", *IBM System Journal*, vol. 42, no. 1, pp. 5–18, 2003.

[GRO 02] GROUP Y., How Much is an Hour of Downtime Worth to You?, Must-know Business Continuity Strategies, pp. 178–187, 2002.

[HOR 01] HORN P., Autonomic Computing: IBM's Perspective on the State of Information Technology, IBM Corporation, 2001.

[HSU 10] HSU I.Y.-Y., WÓDCZAK M., WHITE R.G. *et al.*, "Challenges, approaches, and solutions in intelligent transportation systems", *2010 Second International Conference on Ubiquitous and Future Networks (ICUFN)*, Jeju island, pp. 366–371, 2010.

[IBM 05] IBM, An architectural blueprint for autonomic computing, Technical report 3rd ed., IBM, Hawthorne, available at: http://www03.ibm.com/autonomic/ pdfs/ACBlueprintWhite PaperV7. Pdf, 2005.

[INS 12] INSAURRALDE C., "Autonomic management for the next generation of autonomous underwater vehicles", *IEEE/OES Autonomous Underwater Vehicles (AUV)*, Southampton, pp. 1–8, 2012.

[JAF 12] JAFFAR S., SUBRAMANYM M.V., "Broadcasting methods in mobile ad hoc networks: taxonomy and current state of the art", *Global Journal of Computer Science and Technology*, vol. 12, no. 1, pp. 59–65, 2012.

[KAM 10] KAMINI R.K., "Vanet parameters and applications: a review", *Global Journal of Computer Science and Technology*, vol. 10, pp. 72–77, 2010.

[KAR 10] KARTHIKEYAN N., PALANISAMY V., DURAISWAMY K., "Optimum density based model for probabilistic flooding protocol in mobile ad hoc network", *European Journal of Scientific Research*, vol. 39, no. 4, pp. 577–588, 2010.

[LI 12] LI J., WÓDCZAK M., WU X. *et al.*, "Vehicular networks and applications: challenges, requirements and service opportunities", *International Conference on Computing, Networking and Communications (ICNC)*, Maui, Hawaii, pp. 660–664, 2012.

[LIM 00] LIM H., KIM, C., "Multicast tree construction and flooding in wireless ad hoc networks", *Proceedings of the 3rd ACM International Workshop on Modeling, analysis and Simulation of Wireless and Mobile Systems*, pp. 61–68, 2000.

[NAN 10] NA NAKORN N., ROJVIBOONCHAI K., "POCA: position-aware reliable broadcasting in VANET", *2nd Asia-Pacific Conference of Information Processing (APCIP)*, pp. 17–18, 2010.

[NGU 07] NGUYEN D., MINET P., "Analysis of MPR selection in the OLSR protocol", *International Conference on Advanced Information Networking and Applications Workshops*, vol. 2, pp. 887–892, 2007.

[NI 99] NI S.-Y., TSENG Y.-C., CHEN Y.-S. *et al.*, "The broadcast storm problem in a mobile ad hoc network", *MobiCom '99: Proceedings of the 5th Annual ACM/IEEE International Conference on Mobile Computing and Networking*, pp. 151–162, 1999.

[PAR 12] PARIDEL K., BALEN J., BERBERS Y. *et al.*, "VVID: a delay tolerant data dissemination architecture for VANETs using V2V and V2I communication", *The Second International Conference on Mobile Services, Resources, and Users, MOBILITY 2012*, pp. 151–156, 2012.

[PEN 00] PENG W., LU X.-C., "On the reduction of broadcast redundancy in mobile *ad hoc* networks", *Proceedings of the 1st ACM International Symposium on Mobile ad hoc Networking & Computing (MobiHoc'00)*, pp. 129–130, 2000.

[ROS 09] ROS F.J., RUIZ P.M., STOJMENOVIC I., "Optimum density based model for probabilistic flooding protocol in mobile ad hoc network", *69th IEEE Vehicular Technology Conference (VTC Spring 2009)*, pp. 1–5, 2009.

[STE 03] STERRITT R., BUSTARD D.W., "Towards an autonomic computing environment", *DEXA Workshops*, Prague, Czech Republic, pp. 699–703, 2003.

[STO 04] STOJMENOVIC T., WU J., "Broadcasting and activity-scheduling in ad hoc networks", in STOJMENOVIC I. (ed.), *Ad Hoc Networking,* IEEE Press, 2004.

[SUT 07] SUTHAPUTCHAKUN C., GANZ A., "Priority Based Intervehicle Communication in Vehicular Ad-hoc Networks Using IEEE 802.11e", *IEEE VTC Spring*, pp. 2595–2599, 2007.

[TON 10] TONGUZ O.K., WISITPONGPHAN N., BAI F., "DV-CAST: a distributed vehicular broadcast protocol for vehicular *ad hoc* networks", *IEEE Wireless Communications*, vol. 17, no. 2, pp. 47–57, 2010.

[VEG 13] VEGNI A.M., BIAGI M., CUSANI R., "Smart vehicles, technologies and main applications in vehicular ad hoc networks", available at: http://www.intechopen.com/books/export/citation/BibTex/vehiculartechnologies-deployment-and-applications/smartvehiclestechnologies - and-main-applications-in-vehicular-ad-hoc-networks, 2013.

[WIL 02] WILLIAMS B., CAMP T., "Comparison of broadcasting techniques for mobile ad hoc networks", *Proceedings of the ACM International Symposium on Mobile Ad Hoc Networking and Computing (MOBIHOC)*, pp. 194–205, 2002.

[WOD 12] WODCZAK M., "Autonomic cooperative networking for vehicular communications", *11th International Conference on Ad-Hoc, Mobile, and Wireless Networks Service, ADHOC-NOW'12*, pp. 112–125, 2012.

[WU 03] WU J., LOU W., "Forward-node-set-based broadcast in clustered mobile *ad hoc* networks", *Wireless Communication and Mobile Computing*, vol. 3, pp. 155–173, 2003.

[YU 06] YU S., CHO G., "A selective flooding method for propagating emergency messages in vehicle safety communications", *2006 International Conference on Hybrid Information Technology*, pp. 556–561, 2006.

List of Authors

Wahabou ABDOU
LE2i
University of Burgundy
Dijon
France

Iñigo ADIN
CEIT and Tecnun
University of Navarra
San Sebástian
Spain

Marina AGUADO
Faculty of Engineering
University of the Basque Country
Bilbao
Spain

Marion BERBINEAU
COSYS department
IFSTTAR
Villeneuve d'Ascq
France

Sébastien BINDEL
MIPS Lab
University of Haute Alsace
Colmar
France

Hervé BOEGLEN
XLIM Lab
University of Poitiers
Futuroscope
France

Guillermo BISTUÉ
CEIT and Tecnun
Univeristy of Navarra
San Sebástian
Spain

Maria CALDERON
Charles III University of Madrid
Leganés
Madrid
Spain

Serge CHAUMETTE
LABRI
University of Bordeaux
Bordeaux
France

Benoit DARTIES
LE2i
University of Burgundy
Dijon
France

Frédéric DROUHIN
MIPS Lab
University of Haute Alsace
Colmar
France

Marco FIORE
Institute of Electronics, Computer and Telecommunication Engineering
Turin
Italy

Fabien GARCIA
Communication Networks Research Group
ENAC
Toulouse
France

Marco GRAMAGLIA
IMDEA Networks Institute
Leganés
Madrid
Spain

Christophe GUERBER
Communication Networks Research Group
ENAC
Toulouse
France

Benoit HILT
MIPS Lab
University of Haute Alsace
Colmar
France

Nader MBAREK
LE2i
University of Burgundy
Dijon
France

Jaizki MENDIZABAL
CEIT and Tecnun
University of Navarra
San Sebástian
Spain

Diala NABOULSI
Concordia University
Montréal
Québec
Canada

Christian PINEDO
Faculty of Engineering
University of the Basque Country
Bilbao
Spain

Alain PIROVANO
Communication Networks Research Group
ENAC
Toulouse
France

Robert PROTZMANN
Automotive Services and Communication Technologies
Fraunhofer Institute for Open Communication Systems
Berlin
Germany

Ilja RADUSCH
Automotive Services and Communication Technologies
Fraunhofer Institute for Open Communication Systems
Berlin
Germany

José RADZIK
Electronic, Optronic, Signal Department
ISAE Supaero
Toulouse
France

Eric RAMAT
LISIC
University of the Littoral Opal Coast
Calais
France

Lara RODRIGUEZ
Faculty of Engineering
University of the Basque Country
Bilbao
Spain

Justinian ROSCA
Corporate Technology
Siemens Corporation
Princeton
USA

Mickaël ROYER
Communication Networks Research Group
ENAC
Toulouse
France

Björn SCHÜNEMANN
Daimler Center for Automotive Information Technology Innovations
Technical University of Berlin
Germany

Patrick SONDI
LISIC
University of the Littoral Opal Coast
Calais
France

Praprut SONGCHITRUKSA
Schlumberger
Houston
USA

Srinivasa SUNKARI
Texas A&M Transportation Institute
College Station
USA

Oscar TRULLOLS-CRUCES
Barcelona Supercomputing Center
Spain

Ines UGALDE
Electrical and Computer Engineering
Rutgers University
Piscataway
USA

Alexey VINEL
School of Information Technology
Halmstad University
Sweden

Index

A, B, C

ad-hoc networks, 1, 2, 25, 51, 54, 58
ADM, 212, 221, 222
air traffic regulations, 56
ATP, 33
automated driving, 150
autonomic manager, 212, 219, 221, 223, 224, 229
autonomous driving, 138, 158, 159
AVLC protocol, 64–66, 70
BER, 43, 44, 64, 108, 111, 113, 115–117, 122–124
broadcast, 1, 3, 7, 14, 17, 68, 72, 74, 141, 216–223, 229, 231
cellular networks, 6
connected mobility, 54, 169
Connected Vehicles (CV), 135, 138, 159
cosimulation, 79, 97, 103, 104

D, E, F

Dedicated Short Range Communications (DSRC), 134
density evaluation, 224
DES model, 42
dynamic window size, 186, 205, 208
ERTMS, 81, 86, 97
ETCS, 33, 80–83, 86, 94, 101
floating car data, 18, 20
Friis, 109, 115, 116, 140

G, I, L

GSM-R, 79, 80, 82, 84–87, 90, 93, 94, 101, 102, 104
infrastructure-to-vehicle (I2V) communication, 159
input traffic feed, 167, 168, 174, 176
intelligent dilemma zone avoidance, 135, 149
IoT, 29, 30, 34
ITS, 1, 2, 165
learning of stochastic channel models, 167
Line of Sight (LOS), 109, 141, 149, 175
link quality assessment, 195, 197, 200, 208
LTE, 7, 12, 14, 43, 46, 86, 101, 102, 138

M, N, O

microscopic traffic simulation, 138
mobility models, 54, 55, 167, 169

multi-estimators, 186, 190, 199, 203
Multiple Domain Simulation, 29
near-field, 29, 30, 33
network simulator, 6, 18,30, 43, 47, 57, 81, 87, 111, 114, 116, 134–136, 166, 223
NFC, 35
North Atlantic Tracks, 59
ns-2, 45, 112
ns-3, 45, 107
OPNET, 89, 93, 97
optimization, 218

P, Q, R

parameter fine-tuning, 170
performance evaluation, 229
piggybacking, 64, 70
priority level, 219, 221, 222, 226, 231
Quality of Service, 211
railway safety, 83
Rayleigh distribution, 110, 141
realistic simulation, 111, 133, 187, 208
realistic wireless channel models, 133
RFID, 34
Rice distribution, 110
riverbed-modeler, 43
routing protocol, 206

S, T, V, W

safety pilot model deployment, 143
satellite, 71
SC, 29
self-management, 219
shadowing, 17, 110
Signal Phase and Timing (SPaT), 135, 150
simulation, 1, 51, 67, 69, 165, 201, 202, 211, 222
simulation framework, 5
simulation of Cooperative Trajectories, 52, 60
smart-flooding, 216, 221, 222, 230, 231
synthetic mobility, 5, 167, 169, 177
telecommunication, 84
topology analysis, 66, 67, 182, 225, 228
trainsignaling, 81
V2X communication, 7, 8, 13–15
VDL card, 67, 68
vehicular ad-hoc networks (VANET), 107
vehicular networks, 178, 185
Wireless Access in Vehicular Environment (WAVE), 108, 110, 165

Other titles from

iSTE

in

Networks and Telecommunications

2017

BENSLAMA Malek, BENSLAMA Achour, ARIS Skander
Quantum Communications in New Telecommunications Systems

2016

AL AGHA Khaldoun, PUJOLLE Guy, ALI-YAHIYA Tara
Mobile and Wireless Networks (Advanced Network Set – Volume 2)

BATTU Daniel
Communication Networks Economy

BENSLAMA Malek, BATATIA Hadj, MESSAI Abderraouf
Transitions from Digital Communications to Quantum Communications: Concepts and Prospects

CHIASSERINI Carla Fabiana, GRIBAUDO Marco, MANINI Daniele
Analytical Modeling of Wireless Communication Systems (Stochastic Models in Computer Science and Telecommunication Networks Set – Volume 1)

EL FALLAH SEGHROUCHNI Amal, ISHIKAWA Fuyuki, HÉRAULT Laurent, TOKUDA Hideyuki
Enablers for Smart Cities

PEREZ André
VoLTE and ViLTE

2015

BENSLAMA Malek, KIAMOUCHE Wassila, BATATIA Hadj
Connections Management Strategies in Satellite Cellular Networks

BENSLAMA Malek, BATATIA Hadj, BOUCENNA Mohamed Lamine
Ad Hoc Networks Telecommunications and Game Theory

BERTHOU Pascal, BAUDOIN Cédric, GAYRAUD Thierry, GINESTE Matthieu
Satellite and Terrestrial Hybrid Networks

CUADRA-SANCHEZ Antonio, ARACIL Javier
Traffic Anomaly Detection

LE RUYET Didier, PISCHELLA Mylène
Digital Communications 1: Source and Channel Coding

PEREZ André
LTE and LTE Advanced: 4G Network Radio Interface

PISCHELLA Mylène, LE RUYET Didier
Digital Communications 2: Digital Modulations

PUJOLLE Guy
Software Networks (Advanced Network Set – Volume 1)

2014

ANJUM Bushra, PERROS Harry
Bandwidth Allocation for Video under Quality of Service Constraints

BATTU Daniel
New Telecom Networks: Enterprises and Security

BEN MAHMOUD Mohamed Slim, GUERBER Christophe, LARRIEU Nicolas, PIROVANO Alain, RADZIK José
Aeronautical Air–Ground Data Link Communications

BITAM Salim, MELLOUK Abdelhamid
Bio-inspired Routing Protocols for Vehicular Ad-Hoc Networks

CAMPISTA Miguel Elias Mitre, RUBINSTEIN Marcelo Gonçalves
Advanced Routing Protocols for Wireless Networks

CHETTO Maryline
Real-time Systems Scheduling 1: Fundamentals
Real-time Systems Scheduling 2: Focuses

EXPOSITO Ernesto, DIOP Codé
Smart SOA Platforms in Cloud Computing Architectures

MELLOUK Abdelhamid, CUADRA-SANCHEZ Antonio
Quality of Experience Engineering for Customer Added Value Services

OTEAFY Sharief M.A., HASSANEIN Hossam S.
Dynamic Wireless Sensor Networks

PEREZ André
Network Security

PERRET Etienne
Radio Frequency Identification and Sensors: From RFID to Chipless RFID

REMY Jean-Gabriel, LETAMENDIA Charlotte
LTE Standards
LTE Services

TANWIR Savera, PERROS Harry
VBR Video Traffic Models

VAN METER Rodney
Quantum Networking

XIONG Kaiqi
Resource Optimization and Security for Cloud Services

2013

ASSING Dominique, CALÉ Stéphane
Mobile Access Safety: Beyond BYOD

BEN MAHMOUD Mohamed Slim, LARRIEU Nicolas, PIROVANO Alain
Risk Propagation Assessment for Network Security: Application to Airport Communication Network Design

BEYLOT André-Luc, LABIOD Houda
Vehicular Networks: Models and Algorithms

BRITO Gabriel M., VELLOSO Pedro Braconnot, MORAES Igor M.
Information-Centric Networks: A New Paradigm for the Internet

BERTIN Emmanuel, CRESPI Noël
Architecture and Governance for Communication Services

DEUFF Dominique, COSQUER Mathilde
User-Centered Agile Method

DUARTE Otto Carlos, PUJOLLE Guy
Virtual Networks: Pluralistic Approach for the Next Generation of Internet

FOWLER Scott A., MELLOUK Abdelhamid, YAMADA Naomi
LTE-Advanced DRX Mechanism for Power Saving

JOBERT Sébastien *et al.*
Synchronous Ethernet and IEEE 1588 in Telecoms: Next Generation Synchronization Networks

MELLOUK Abdelhamid, HOCEINI Said, TRAN Hai Anh
Quality-of-Experience for Multimedia: Application to Content Delivery Network Architecture

NAIT-SIDI-MOH Ahmed, BAKHOUYA Mohamed, GABER Jaafar, WACK Maxime
Geopositioning and Mobility

PEREZ André
Voice over LTE: EPS and IMS Networks

2012

AL AGHA Khaldoun
Network Coding

BOUCHET Olivier
Wireless Optical Communications

DECREUSEFOND Laurent, MOYAL Pascal
Stochastic Modeling and Analysis of Telecoms Networks

DUFOUR Jean-Yves
Intelligent Video Surveillance Systems

EXPOSITO Ernesto
Advanced Transport Protocols: Designing the Next Generation

JUMIRA Oswald, ZEADALLY Sherali
Energy Efficiency in Wireless Networks

KRIEF Francine
Green Networking

PEREZ André
Mobile Networks Architecture

2011

BONALD Thomas, FEUILLET Mathieu
Network Performance Analysis

CARBOU Romain, DIAZ Michel, EXPOSITO Ernesto, ROMAN Rodrigo
Digital Home Networking

CHABANNE Hervé, URIEN Pascal, SUSINI Jean-Ferdinand
RFID and the Internet of Things

GARDUNO David, DIAZ Michel
Communicating Systems with UML 2: Modeling and Analysis of Network Protocols

LAHEURTE Jean-Marc
Compact Antennas for Wireless Communications and Terminals: Theory and Design

RÉMY Jean-Gabriel, LETAMENDIA Charlotte
Home Area Networks and IPTV

PALICOT Jacques
Radio Engineering: From Software Radio to Cognitive Radio

PEREZ André
IP, Ethernet and MPLS Networks: Resource and Fault Management

TOUTAIN Laurent, MINABURO Ana
Local Networks and the Internet: From Protocols to Interconnection

2010

CHAOUCHI Hakima
The Internet of Things

FRIKHA Mounir
Ad Hoc Networks: Routing, QoS and Optimization

KRIEF Francine
Communicating Embedded Systems / Network Applications

2009

CHAOUCHI Hakima, MAKNAVICIUS Maryline
Wireless and Mobile Network Security

VIVIER Emmanuelle
Radio Resources Management in WiMAX

2008

CHADUC Jean-Marc, POGOREL Gérard
The Radio Spectrum

GAÏTI Dominique
Autonomic Networks

LABIOD Houda
Wireless Ad Hoc and Sensor Networks

LECOY Pierre
Fiber-optic Communications

MELLOUK Abdelhamid
End-to-End Quality of Service Engineering in Next Generation Heterogeneous Networks

PAGANI Pascal *et al.*
Ultra-wideband Radio Propagation Channel

2007

BENSLIMANE Abderrahim
Multimedia Multicast on the Internet

PUJOLLE Guy
Management, Control and Evolution of IP Networks

SANCHEZ Javier, THIOUNE Mamadou
UMTS

VIVIER Guillaume
Reconfigurable Mobile Radio Systems